# THE FOUNTAIN OF KNOWLEDGE

INNOVATION AND TECHNOLOGY IN THE WORLD ECONOMY

MARTIN KENNEY, *Editor*
*University of California, Davis, and Berkeley Roundtable*
*on the International Economy*

*Other titles in the series:*

# The Fountain of Knowledge

## THE ROLE OF UNIVERSITIES
## IN ECONOMIC DEVELOPMENT

*Shiri M. Breznitz*

STANFORD BUSINESS BOOKS

*An Imprint of Stanford University Press*

*Stanford, California*

Stanford University Press
Stanford, California

Special discounts for bulk quantities of Stanford Business Books
are available to corporations, professional associations, and other
organizations. For details and discount information, contact the
special sales department of Stanford University Press.
Tel: (650) 736-1782, Fax: (650) 736-1784

Printed in the United States of America on acid-free, archival-quality
paper

Library of Congress Cataloging-in-Publication Data

Breznitz, Shiri M., author.
   The fountain of knowledge : the role of universities in economic
development / Shiri M. Breznitz.
      pages cm — (Innovation and technology in the world economy)
   Includes bibliographical references and index.
   ISBN 978-0-8047-8961-5 (cloth : alk. paper)
      1. Universities and colleges—Economic aspects—Case studies.
2. Technology transfer—Case studies.   3. Biotechnology—Technology
transfer—Case studies.   4. Academic-industrial collaboration—
Economic aspects—Case studies.   5. Yale University.   6. University
of Cambridge.   I. Title.   II. Series: Innovation and technology in
the world economy.
   LC67.6.B72 2014
   338.4'3378—dc23
                                                   2013049697

   ISBN 978-0-8047-9192-2 (electronic)

Typeset by Newgen in 10/13 Galliard

*To Danny, the wind beneath my wings*

# Contents

# Tables, Figures, and Map

MAP

*Preface*

*The Fountain of Knowledge* is the story of the role of the university in the changing world, and in particular its role in economic development. It is not a book about technology transfer or a book about organizational change. It is not a book about the biotechnology industry. It is also not the story of Yale and Cambridge (or Cambridge and Yale—making sure no one gets insulted). All of the above are important parts that allow me to tell a story. But it is a different story.

Early in my academic career, I studied the localization of the biotechnology industry in Massachusetts. I asked, why does the biotechnology industry in Massachusetts cluster in Cambridge rather than in Boston or any other town? What I found was that it all was connected to MIT and Harvard. I was intrigued, and that is what led me to ask the next question, why universities?

For most of us, universities are needed for getting an education, getting a good job, and maybe moving up the professional ladder. We see universities as something we need to have. But do we? A growing number of policy makers, university presidents, and evidence in the form of economic growth say we do. Since the early nineteenth century there has been a growing pressure on universities to be participating citizens and contribute to society. But with the high-tech boom and the success of many university-related firms, such as HP, Google, and Yahoo, policy makers have started to view universities as the pathway to a region becoming a successful part of a global economy.

When we examine some of the most economically successful regions in the world, when we try to understand what is behind their success, we find a university. A university—really?

Regions that are trying to change and improve their economic base look to Cambridge, Massachusetts; Cambridge in the United Kingdom; and Helsinki and Turku in Finland, and what do they find? Many small to medium-size enterprises. Many of these were spun out or have some relationship with the local university or universities. Should universities, which were created to support the unbounded world of ideas, focus on applied research? *The Fountain of Knowledge* is a book about universities and their role in today's global economy: how (and why) they are contributing to their local economies.

Why did I choose to study Yale University and the University of Cambridge? The answer is simple. Both are world-renowned research universities that made an effort to change and make an economic contribution to their local regions. I was in the right place at the right time—I was at each of these institutions immediately after they began the process toward change, and during the process. This is a rare opportunity, and I took advantage of it. I conducted many interviews with faculty, researchers, companies, members of the business community, and local government representatives, all dealing with the question of how to maximize revenues (though to each person I interviewed, "revenues" meant something else).

My choice to study biotechnology was also simple. Biotech is a relatively new technology that, unlike many others, requires a close relationship with a research laboratory, in many cases within universities, hospitals, or research institutes. It is fairly easy to map the relationships of firms to their origin universities. Many executives of biotechnology firms are former PhD students who had a keen interest in my work and were willing to be interviewed. Without them this book would not exist.

I believe that universities should be an invaluable factor for any economic development studies. Hence, in this book I analyze universities in a wide scope, including the history and environment of the region in which universities operate. I draw much from organizational studies, but I view technology commercialization and knowledge transfer as more important in shedding light on how and why these changes occur and which worked best. Hence, I analyze the focus on technology commercialization in lieu of its impact on the local economy. Last, while conducting this research, and unlike previous researchers, I did not choose cases by their success. Instead, I followed two universities over time and evaluated the impact of their efforts to improve technology commercialization on the local economy.

# *Acknowledgments*

This book would have not been written if it were not for the help of a few remarkable people. At each university studied there were people willing to explain procedures, provide data, and answer my many calls and requests for interviews. I especially want to thank Jonathan Soderstrom at Yale University and David Secher and Richard Jennings at the University of Cambridge. Moreover, friends and mentors continuously helped along the way—I would like to thank Mia Gray for all the hard work, her patience, her encouragement, and insights. The MIT's Industrial Performance Center team, and in particular professors Richard Lester and Suzanne Berger, provided guidance and support even before this research project started. Maryann Feldman taught me new methods and guided me through the academic maze. Margo Beth Fleming and Martin Kenney at Stanford University Press provided important comments and new perspectives that brought this manuscript to its current state. Sharmistha Bagchi-Sen, Henry Etzkowitz, Laurent Frideres, Joyce Ippolito, Helen Lawton Smith, Jason Owen-Smith, John Walsh, Lynne Zucker, and John Zysman provided much-needed advice and support. The encouragement of my wonderful family supported me along the way: my parents, Mordehay and Gina Marom, raised me to believe that I can do anything I set my mind to. With their support I always feel that anything is possible; my beloved sister Rinat had saved me from my own crazy ideas and provided me with a relaxed and safe haven both physically and emotionally; my nephews Ben and Matan, and my niece Adi, always lifted my spirit; Tammy's support and her belief in me were remarkable not only for a friend but, and especially so, for a mother-in-law. Danny, Mika, and Tom, you are the world to me. This book represents my work and our life in one. I know exactly which chapter I was writing when Mika was born and the specific revision I was working on when Tom came along. Danny was my second adviser, and a true partner for the road. I am blessed to have you in my life. This book is as much yours as it is mine.

# THE FOUNTAIN OF KNOWLEDGE

## CHAPTER ONE

# Introduction

Universities, viewed as fountains of knowledge, produce the world's most important resources: young minds and an educated labor force, which in turn produce cutting-edge research and innovative ideas and products that contribute directly to economic development. Thus, universities contribute directly to a region's economic growth, making universities a highly desirable and almost essential resource for a region.

The economic development effected by a university is evaluated by the amount of technology commercialization it generates. Patents, licenses, and spin-out firms are easy to quantify and use as a measurement of university productivity. Technology firms tend to develop near universities as a result of the knowledge spillover generated by university research. The mere existence of a university in a region, however, is not a guarantee of economic success.

So what determines how well a region benefits from the presence of an innovative research university? There are two main factors: the ability of the university to transfer knowledge to the public domain and to commercialize technology, and the region's ability to absorb that information. Some regions are better able than others to innovate and commercialize technology. By focusing on university knowledge transfer and technology commercialization, the fountain of knowledge, we can evaluate whether the research that universities are expected to do is beneficial and valuable to local economic development—after all, it is demanding and requires many resources of these higher education institutions.

*The Fountain of Knowledge* analyzes two world-renowned universities, their investment in technology commercialization, and the outcome of that contribution in their local economies. This book has three main arguments. One is that the way in which a university goes about improving its technology-transfer capability matters. Conducting a focused and thoughtful comprehensive change that includes all sections of the university will improve commercialization. Second, by choosing a particular path of change, the university also changes its role and its ability to contribute to the region. Third, not all changes will positively affect the local economy.

## THE ROLES OF THE UNIVERSITY

The evaluation of university technology commercialization is vital, considering that the traditional roles of universities have been research and teaching. These original roles of universities have been intensely discussed over the past century at both national and regional levels. Should universities remain islands of research, free of politics, economics, and social class? Or should they participate as active players in local economies and societies?

Historically, universities were the domain of the upper classes, who studied such esoteric subjects as literature and philosophy. Over time, universities began to serve the general public, offering more practical subjects, such as applied research, and training students for professions like medicine and law. By the early 1900s universities had become recognized as regional and national engines of growth.

The modern university, as it developed in the nineteenth century, is an important source of new knowledge and technologies, with the potential to be commercialized (Scott 1977). Today's model of the university has a public-service component, offering a wider base for research and teaching—both of which have the power to promote social change. According to Scott (1977), the service component was a direct result of changes in modern society—that is, growth in the number of students and demand for skilled workers. The university service component was influenced by a neoliberal economic perspective. From that perspective, universities are evaluated on the basis of their contribution to the economy. Therefore, in most countries, universities that rely heavily on public funding are pressured to "pay back" the community and act like responsible citizens (Russell 1993).

The pressure on modern universities to pay back the community has created what is known as the "third role" of universities. Many universities are now obliged to make a contribution to society through research and development (R&D), collaborations, and technology transfer with industry (Minshall, Druilhe, and Probert 2004). Collaborating with industry is a significant change from the original mission of the university, representing an expectation of service that many institutions are not ready or willing to make. However, there is an apparent public benefit for industry collaborations. Universities are an important source of a skilled labor force that is often trained through public funding. Moreover, commercialization can be a solution for universities' financial constraints as well as a way for students to gain industry experience. Hence, university-industry collaboration and proximity promote the formation of industry and economic growth.

However, there is still a debate over university-industry relationships. Studies of higher education institutions emphasize the ability of universities to become important contributors because they are centers of "free" thinking. The idea behind the tenure-track position was to allow faculty to work on new ideas without any constraints. Some of the most interesting innovations, such as electricity, started with a totally different research question in mind. In many cases the discoveries were even found in different schools of thought. Should we predicate what universities need to work on because we know they are capable of producing the next generation of technology? Are we not limiting the fountain of knowledge this way to a mere drizzle?

Some scholars believe that there should be a separation of university and industry. Those scholars claim that academics do not possess the business knowledge to determine which projects should be commercialized, nor should public universities provide services to a specific private market or a particular industry. Technology transfer requires universities to be attuned to, and work with, industry's business perspective. Spin-out companies in particular require different business skills from those that universities are normally equipped with, such as expertise in entrepreneurship, business development, and venture capital. Furthermore, technology-transfer offices need to be able to assess inventions and decide whether they have a commercial value. Thus, they need to employ specialists or hire consultants to evaluate the technology, which is often expensive. Is this the knowledge we are looking for when we consider universities as fountains of knowledge?

## TO MEASURE?

Many claim that by being fountains of knowledge, universities already contribute to regional and national economies. Hence, their contribution cannot be ignored and should be used to improve the quality of life and the economic situation of the region and community in which the university is located. There are two ways to measure the impact of a university on a regional economy. The classic, or "short run," method is to determine the institution's contribution to the annual flow of regional economic activity. The "long-run" method focuses on the contribution of the institution to the continuous growth of human capital in the region (Beck et al. 1995).

The short-run impact—actual dollars flowing into a region due to the mere presence of the university—can be measured by the purchases made by the university in the region: office supplies, rent, food, and services; salaries for employees, some of whom live in the region and spend their wages in the region. Outside funds like donations, grants, and state and federal government funding to the university are also considered in determining a university's economic impact. In this way a university is measured only by direct input and output. University contributions that are not measured in dollar amounts, such as graduate students' firms and firms' products based on university research, are not taken into consideration.

The long-run impact measures "the future income stream of graduates who stay to work in the area" (Beck et al. 1995, p. 246) and the economic impact of graduate students' firms and firms with products based on university research and patents. Measuring long-run impact provides a method to calculate the return on tax invested in higher education. Studies have proved that higher education leads to higher levels of income. In urban areas, the presence of universities seems to affect the growth rates, earnings, and composition of employment. Hence, the ability of a university to patent, license, and spin out firms has a direct impact on the long-run economic development of a region, which is the focus of this book.

## THIS BOOK

This book examines the cases of two prominent universities, Yale University and the University of Cambridge, that have made policy, culture, and organizational changes to improve their abilities to commercialize technologies and to have a wider impact on their respective local economies. Interestingly, these two universities took different approaches to technology transfer and had different outcomes, both for themselves and for their local regions. Though previous studies have found university investment in technology commercialization to have a positive impact, I found that not all results have been positive.

I also consider other factors that may have been responsible for the changes at both locations. For example, a university's ability to disseminate academic ideas to the private market and to contribute to regional economic growth frequently depends on internal mechanisms and resources rather than on formulaic technology-transfer models.

Moreover, universities do not exist in a vacuum—they are influenced by social and economic processes and politics. Each university should be analyzed in its historical and environmental arena. This kind of analysis is a more valid indicator as to the ability of a university to disseminate academic ideas to the private market and to commercialize inventions.

I have analyzed the two universities in lieu of the local biotechnology industry. This industry relies heavily on university research. As a result, ties between biotechnology companies and specific research institutes are easy to identify. Biotechnology is a "new" industry, with its earliest companies established in the 1970s,[1] but biotechnology itself is not a new phenomenon. What we know as modern was the result of several breakthroughs in molecular biology during the mid-twentieth century (Acharya 1999). In 1953, James Watson and Francis Crick, from the University of Cambridge, identified the structure of DNA. This breakthrough was followed by the development of monoclonal antibodies, on which diagnostic kits in the therapeutic industry are based. First developed at University of California–San Francisco and Stanford University in 1973, the process of cutting and rejoining DNA to produce recombinant DNA that could replicate a host cell—known as cloning—revolutionized modern biotechnology.

Research at universities has led to the identification of many new antibodies, proteins, and potential drugs. In many respects, the biotechnology industry has been launched from universities and research institutes, thus creating a clear and direct connection between biotech industries and universities, and providing an excellent case study for this book. While university inventions are usually licensed to private companies, the companies stay in constant contact with the university researchers, especially in their early stages of research, to assist in product development (Kenney 1986).[2] Therefore, the biotechnology industry has been instrumental in the renewal of interest in university-industry relationships and in the commercialization potential of university research (Blankenburg 1998).

Both Yale and Cambridge have been cited as strong research universities in life sciences, instrumental to the development of regional biotechnology clusters. Yale is situated in New Haven County, which has the largest biotechnology industry agglomeration in Connecticut. The biotechnology industry in New Haven, which in 1993 consisted of only of six companies, had grown to forty-nine by 2004 and seventy firms by 2013, making New Haven seventh in the United States by number of biotech companies per capita and third in the nation by per capita research grants (Ernst & Young 2001; US Department of Health and Human Services and National Institutes of Health and US Census 2008). The majority of the biotechnology companies in this cluster have spun out of Yale University. Similarly, the University of Cambridge is located within Cambridgeshire County, with approximately 154 biotechnology companies in this county, representing about a third of all UK biotechnology industry (Sainsbury 1999; Greater Cambridge Greater Peterborough Enterprise Partnership 2013).

Despite its strength in the life sciences, Yale University did not promote technology transfer and commercialization until the mid-1990s. Moreover, its disapproving attitude toward applied research caused the university to lose many faculty members and patents to other universities. However, in 1993, with a change of leadership and concerns for the university's eminence, Yale started to invest in technology transfer and local economic growth, and the investment paid off handsomely. Similarly, in the late 1990s, as a result of government pressure, the University of Cambridge made changes and investments in its technology-transfer policy. But, as we shall see, the university-industry relationships that were formed near Cambridge did not result in revenues for the university, nor did they lead to regional economic growth.

Why such different outcomes? Both institutions are distinguished universities conducting world-class research in the biosciences, and both decided by the late 1990s to make changes to their technology transfer mechanisms. The ways they implemented those changes were very different, however, and as a result, the changes produced significantly different outcomes. As the goal here is to investigate universities' technology-transfer capability, rather than compare the two university cases by output, this book looks at their unique policies, organizational structure, and commercialization culture, all of which have influenced their technology-transfer capabilities.

## THE LITERATURE DEBATE

Although many studies treat technology transfer as a single process rather than focusing specifically on policy and organization, the studies can be divided into two categories. First, some scholars view the university technology-transfer process, including organization, as a factor in the university's ability to commercialize innovative ideas (Feldman and Desrochers 2003; Siegel, Waldman, and Link 2003a; Rothaermel, Agung, and Jiang 2007). Other scholars focus specifically on university technology-transfer output in the form of patents, licenses, and spin-outs (Di Gregorio and Shane 2003; Mowery and Sampat 2001a; Shane 2004).

In terms of the impact of technology transfer on the university itself, the role of the university as a public, nonprofit educational institution is becoming blurred. Universities are now considered leading players in today's global economy, players that can promote and establish certain regions as leaders of the world's economy. Thus, universities are being examined and pressured to prove their ability to innovate as well as to transfer their innovation to the public domain. By allowing universities to own and commercialize their innovations, their success is now measured by a new factor—technology commercialization.

This book continues the investigative tradition of the scholars who believe that to understand the ability of a university to transfer academic ideas to the private market, we must understand technology commercialization investment and processes as a whole. However, unlike other studies that analyzed one or two universities and reviewed one or two factors that affect technology commercialization (e.g., policy, employee characteristics), this book reviews factors identified by previous studies in a comparative study that adds the environment and history of a university location to its analysis. Technology commercialization functions differently in different universities, and as such it has been proved to make a difference in a university's ability to patent, license, and spin out (Feldman et al. 2002; Kenney and Goe 2004; Shane 2004; Siegel, Waldman, and Link 2003a, 2003b). In most studies, universities' investments in technology transfer and commercialization are viewed positively. Interestingly, studies show that although many universities invest heavily in technology transfer, not all show an increase in output as a contribution to their local economy.

## A TALE OF TWO UNIVERSITIES

Enter the arena two of the world's most renowned universities: the University of Cambridge in the United Kingdom and Yale University in the United States. During the 1990s both universities found themselves under pressure to make an impact and contribution to their local economies.

Both institutions made a vast change in their policies and processes to be able to do so.

More specifically, the University of Cambridge, which had strong university-industry relationships and a large number of university spin-outs, executed changes to its policy and organization without consideration as to how those changes would affect other regional players. Hence, the implementation of these changes damaged its technology-transfer capability, as was evident in the reduction of its spin-outs and the response of the local industry. In contrast, Yale University implemented different policy and organization changes while collaborating with other regional players and local industry. The result was a positive impact, evident in the growth of university spin-out and local industry response.

Importantly, the difference between the changes undergone by these two universities lies in the process of change. Yale made comprehensive changes to the entire university's approach to technology transfer that included policy, culture, and organization; the University of Cambridge made partial changes to some of its intellectual property rights policy and organization. Moreover, the velocity of change was different. Yale made a decision to change and started one process that took three years. Cambridge, conversely, made incremental changes to its policy and organization over a period of eight years. Last, while Yale made a conscious decision to include the local industry and region in its changes, Cambridge made changes within the university without input from or cooperation with the local industry or region. Those changes worked out in different ways for the regions, for the firms and financiers in them, and for the universities.

### The University of Cambridge

The mission of the University of Cambridge is to contribute to society through the pursuit of education, learning, and research at the highest international levels of excellence. (University of Cambridge 2005b)

As one of the leading universities in the world, the University of Cambridge is strong in the sciences, specifically in engineering and biomedicine (*Times Higher Education* 2013). It was established in 1209 by scholars who had left Oxford University. The first college, Peterhouse, was established in 1284, creating the foundation for the collegiate system—a defining feature of the university.

Today, Cambridge is a complex web of independent departments, colleges, and research institutes, where central university administration is weak compared to that of other universities in Europe and the United States.[3] Students and most faculty members belong to both a department and a college. The colleges are autonomous institutes that select their own faculty and students, although they are connected to the university through membership in the university council and representation on different boards. Thus, college faculty do not necessarily have positions in a university department, and university department faculty do not necessarily have to hold a position at a college. Students, however, always belong to both the university and a college. The colleges differ by financial capability, educational strengths, and students. Three accept only women; most accept both undergraduates and graduate students. The financial capability of the colleges, including the ownership of land, has contributed to the development of the high-tech industry in the region through the creation of Cambridge Science Park by Trinity College and the St. John Innovation Centre by St. John's College (Gray and Damery 2004).

Cambridge is also known for its relaxed, noncontrolling "policy toward commercial exploitation of academic know-how and links with industry generally" (Segal Quince Wicksteed 1985, 47). The authors of the *Cambridge Phenomenon* are referring here to the free hand that academics are given in Cambridge regarding the commercialization of their research. In applied sciences, it is presumed that faculty will be involved in consulting and research with industry. Thus, it is not surprising to find a growing biotechnology cluster in the region. However, a close examination of the 153 biotechnology companies in Cambridgeshire County found that only 16 percent of the companies are spin-outs from the University of Cambridge (compared to 40 percent of the companies in New Haven).

Figure 1.1 illustrates the constant growth in the number of patents applied for by the University of Cambridge. These figures reinforce data on the academic strength of Cambridge, ranking the university as one of the top patent owners, along with Massachusetts Institute of Technology (MIT), California Institute of Technology, Stanford University, and Johns Hopkins University (US Patent and Trademark Office 2003).[4]

Unlike many other top patent-owning universities, Cambridge is a public university, heavily dependent on government funding. From 1999 to 2012, the University of Cambridge's income grew by 186 percent (from £293 million to £860 million), including an increase of 192 percent in income from grants and contracts (*Times Higher Education* 1999). An indicator of the university's pledge to academic excellence in the biomedical field is the fact that half of funding is dedicated to clinical medicine and

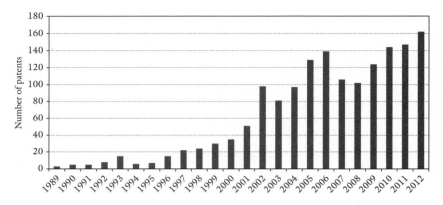

FIGURE I.I University of Cambridge by patents, 1989–2012
SOURCE: European Patent Office (2013).
NOTE: According to this figure, the University of Cambridge has applied for patents on a growing basis since 1989.

TABLE I.I Incomes of the University of Cambridge, 2011–12

| Funding source | Annual income (thousands) | Percentage of total income |
|---|---|---|
| Higher Education Funding Council for England and the Training and Development Agency for Schools grants | £197,265 | 23.5 |
| Research grants and contracts | £293,441 | 34.9 |
| Fee income | £149,234 | 17.7 |
| Endowment and investment income | £54,383 | 6.5 |
| Other income | £146,438 | 17.4 |
| **Total income** | **£840,761** | |

SOURCE: University of Cambridge (2013).
NOTE: Government funding for UK higher education has grown steadily since 1994. However, the level of funding per student has fallen dramatically.

biosciences. Table I.I shows data on the breakdown of the university's income; 24 percent is based on government funding, and another 35 percent is based on other funding for research grants and contracts.[5] It is interesting to note that total government funding for UK higher education has grown steadily since 1994. However, the level of funding per student has fallen dramatically. With the growth in the number of universities, the number of participating students grew. Thus, funding per university declined. This is the case with Cambridge, where the current endowment of £3 billion ($4.65 billion) is large compared to that of other UK universities but small when compared to that of Harvard, which has a $30.4 billion endowment, and Yale, which has $19.3 billion (Staley 2013).

In the academic year 2011–12, 57 percent of the University of Cambridge's students were in the sciences and 5 percent in the medical school (compared with 13 and 4 percent, respectively, at Yale).[6] Of the faculty, 35 percent are in science, and 3 percent are in the medical school (compared with 11 percent and 53 percent, respectively, at Yale), with a ratio of 6.3 students per faculty member in Cambridge (compared with 3 at Yale and 8 at MIT).[7] On the one hand, this ratio implies that Cambridge's faculty members have less time to invest in their students, work on their own research, and collaborate with other researchers or industries. On the other hand, this ratio exemplifies the importance of teaching to the university's culture. Cambridge is well known for its small-group teaching. Faculty are required to tutor students in small groups of two or three students at a time, which places even more teaching responsibilities on the faculty.

The University of Cambridge sees itself as a national and international university, not as a regional leader. O'Shea et al. (2005) argue that a university's history and resources influence its mission and organization, and also influence the university's knowledge-transfer capabilities. The organization's cultural base, influenced by the organization's history and the history of the decision makers in the organization, affects the way in which the organization makes decisions about issues such as strategy, outlook, and cooperation with other players in the local economy (Schoenberger 1997).

Thus, the fact that Cambridge does not see itself as a regional player influences the way in which the university engages and contributes to the local economy. Therefore, to encourage local academic entrepreneurship, there is a need for a commercially supportive culture at the university. The way Cambridge perceives itself is reflected in its mission "to contribute to Society through the pursuit of education, learning, and research." This also reflects the way in which the university structures university-industry relationships in general and technology transfer in particular.

### Yale University

Each college surrounds a courtyard and occupies up to a full city block, providing a congenial community where residents live, eat, socialize, and pursue a variety of academic and extracurricular activities. Each college has a master and dean, as well as a number of resident faculty members known as fellows, and each has its own dining hall, library, seminar rooms, recreation lounges, and other facilities. (Yale University 2013)

Yale University is the third-oldest institution of higher learning in the United States. Yale was built in 1701 and was renamed Yale College in 1718 after Elihu Yale, who made significant donations to the college. In the 1930s Yale established residential colleges, similar to those at Oxford and the University of Cambridge. Its new model was a distinctive system that divided the undergraduate population into twelve separate communities of approximately 450 members each, allowing Yale to offer both the intimacy of a small college environment and the resources of a major research university. In fall 2011, there were 11,875 students at Yale, of which 17 percent were international students representing 118 countries.

Although Yale is known for its excellence in many fields, including life sciences, Yale University's historical culture of noninvolvement in the community in general and with industry in particular created a situation in which it failed to reap the credit for several important discoveries, such as the transgenic mouse.[8] For many years Yale was not active in technology commercialization, and by 1993 it had spun out only three biotechnology companies. This attitude of noninvolvement in industry changed during 1993–96. On the academic level, in 1994 Yale invested heavily in the life sciences. Out of a total of 729 tenured faculty members, 38 percent (279) taught in the medical school and another 5 percent (36) in the biological sciences (Office of Institutional Research 2001). According to Yale's 1995–96 financial report, income from research grants and contracts represented 29 percent of total income and totaled $262.2 million in fiscal year 1996. Of those funds, nearly 75 percent supported programs in the medical school and the Departments of Biological and Physical Sciences and Engineering. Of the $262.2 million, $203.6 million represented federal government funds, of which 80 percent was awarded by the National Institutes of Health.

However, Yale was a promoter neither of applicable research nor of working with industry. Hence, in 1994, Yale spent $224,939,000 on R&D but registered only sixteen patents. It is interesting to compare these figures with those of MIT, which spent $374,768,000 on R&D in that year and registered ninety-nine patents (National Science Foundation 2003). While Yale spent $14,058,388 per patent, MIT spent $3,785,535 per patent, a ratio almost four times higher for Yale. Also, by 1993, Yale had spun out just three biotech companies, compared with MIT, which had spun out thirty, and only one of Yale's spinouts, Alexion Pharmaceuticals, had stayed in the New Haven region. These figures are broadly consistent with the reputation of Yale at the time as an institution only peripherally and sporadically involved with the local economy and community. As Yale's president Richard Levin noted years later:

> Outsiders have long regarded the presence of Yale as one of the city's major assets, but, except for episodic engagement, the University's contributions to the community did not derive from an active, conscious strategy of urban citizenship. It is true that our students, for more than a century, have played a highly constructive role as volunteers. Even a decade ago, two thousand students volunteered regularly in schools, community centers, churches, soup kitchens, and homeless shelters, but these volunteer efforts were neither coordinated nor well supported institutionally. When I became president, in 1993, there was much to be done to transform Yale into an active, contributing institutional citizen. . . . In prior years, however, the university had taken a relatively passive attitude toward the commercialization of its science and technology. (Yale Office of Public Affairs 2003)

With the exception of faculty in a few departments, such as pharmacology, during this period, Yale faculty members were not encouraged to work on research with practical applications. It was actually implied that the outcome of such involvement would have an unfavorable impact on one's academic career. A former faculty member at Yale during the late 1960s observed: "One of the things that depressed me was that they did not want to do any application. You could consult but that was not a good status" (interview with former Yale faculty member). So even though important discoveries were made at Yale during that period, the Office of Cooperative Research had a somewhat passive attitude toward commercialization, and only a few discoveries were patented.[9] According to another former faculty member:

> [There was] very little applied research in biology, maybe in the medical school or Pharmacology and Chemistry Departments. In the Biology Department it was looked down upon. For example, we made the first transgenic mouse, and [the Office of Cooperative Research] considered that not to be worthwhile in terms of invention. Yale was very conservative for many years. Not a very active program. Yale actually lost a lot of intellectual property because of this culture. They did not patent on time. (Interview with Yale faculty member)

Although only a few biotech firms established themselves prior to 1993, this was not the result of an inhospitable environment. In fact, by 1993, Connecticut had hosted five pharmaceutical companies: Pfizer,

Bristol-Myers Squibb, Purdue, Bayer, and Boehringer Ingelheim. Most of these companies had a major presence in the state, including research facilities; four of these companies were located in the New Haven metropolitan area. In 1995, a total of $1.2 billion was spent on pharmaceutical R&D in Connecticut itself (6 percent of the nation's total). The companies operated research-oriented facilities staffed with scientists with an intensive knowledge base in biomedicine, but interactions with researchers at Yale and other local universities were limited. None of these companies established institutional relationships with local research institutes, relying instead on opportunistic interactions between their investigators and individual researchers at the institutes.

In summary, when we examine these two universities, University of Cambridge and Yale University, we find that while they seem similar on many levels, they are different in their abilities to commercialize technology. We have a public versus private university, an institution that had a history of university-industry relationships and one that did not, one that has funding and allows faculty to focus on research and one that focuses on teaching and has limited research funding. This is the basis for the differentiation of the two universities, but it is also the basis from which we start our journey toward understanding the optimal formula for technology commercialization and knowledge transfer.

## NEXT CHAPTERS

The book is organized as follows: Chapter 2 examines existing studies on university technology commercialization and organizational change, finding that current studies do not entirely explain the differences in universities' ability to commercialize technology. Chapter 3 explains the national framework in which each university has been operating, and in particular focuses on science and technology policy as well as on university system organization and funding in each country. Chapters 4 and 5 provide detailed overviews of the history and organizational change that both Cambridge and Yale experienced, including the role of the universities in the region before and after the changes. Specifically, these chapters provide a detailed explanation of the successful change at Yale University. The book concludes with three lessons. First, knowledge transfer and technology commercialization require the collaboration of many regional and national players. Second, historical and environmental factors have enormous, direct impact on the ability of an institution to commercialize technology. Third, intrauniversity factors, which are highly diverse, are the most important factors to consider in technology commercialization. This research adds three additional factors that affect successful technology commercialization in particular to universities that are trying to improve their technology-transfer process: the velocity of the university's organizational change in relation to commercialization, the level of the change within the university (partial or comprehensive), and the community of change (whether a change is done in collaboration with other players in the region).

## CHAPTER TWO

# Factors Affecting University Technology Transfer

The university is generally seen as making a positive contribution to local economic development in general and entrepreneurship and spin-out formation in particular (Kenney 1986; Zucker, Darby, and Peng 1998; Di Gregorio and Shane 2003; Shane 2004; Goldstein and Renault 2004; Lane and Johnstone 2012). Both external and internal factors affect a university's success in technology transfer (Cooke 2002; Etzkowitz 1995; Kenney and Goe 2004; O'Shea et al. 2005; Siegel, Waldman, and Link 2003a; Boucher, Conway, and Van Der Meer 2003; Bagchi-Sen and Lawton Smith 2012). See Table 2.1.

Although universities have no control over their historical and locational factors—that is, external factors—they do have some level of control over their internal programs, resources, and procedures for transferring technologies. Hence, a study of the technology-transfer activities of universities needs to focus on this area.

Two external factors affect the ability of a university to commercialize technology: history and environment. Historical factors, based on national, international, and regional policies such as intellectual property rights laws and tax incentives, play an important role in the ability of universities to succeed in their technology transfer and university-industry relationships (Lawton Smith and Ho 2006; Mowery et al. 1999; Rahm, Kirkland, and Bozeman 2000; O'Shea et al. 2005).

Environmental factors relate to the relationships among institutions on national and regional levels. The ability of a group of local institutions to transfer knowledge and hence to affect the ability of a locality to inno-

TABLE 2.1 Factors affecting the university's commercialization ability

---

**EXTERNAL FACTORS**

- Legislation, such as the Bayh-Dole Act in the United States, which deals with intellectual property arising from government-funded research. There is specific state legislation as well; California, Massachusetts, and North Carolina have specific tax incentives for the high-tech industry. UK laws allow a small or medium-size enterprise to claim 150% tax relief on R&D, whereas large corporations can claim 125% tax relief.
- The relationships between firms and institutions on national and regional levels. The size and number of players in each location have a direct impact on the national and regional systems of innovation.

---

**INTERNAL FACTORS**

- Cultural factors, such as the innovation and commercialization culture within each university and within departments
- University policies relating to intellectual property rights, division of royalties, and equity policy
- The organization of the university's TTO; TTOs vary widely in terms of quality and amount of personnel, business experience, and past success commercializing technology

---

SOURCE: Etzknowitz (1995); Cooke (2002); Boucher, Conway, and Van Der Meer (2003); Siegel, Waldman, and Link (2003a, 2003b); Kenney and Goe (2004); O'Shea et al. (2005); Bagchi-Sen and Lawton Smith (2012).

vate depends on their number, strength, and collaboration efforts. Sharing of information and collaboration among institutions drives innovation (Nelson 1993).

Internal factors, such as the university's technology-transfer culture, policy, and organization, are built on academic prestige, funding availability, and networking (Shane 2004; Roberts 1991; Zucker, Darby, and Peng 1998; Clark 1998; Lockett and Wright 2005; Kenney and Goe 2004). The institutions considered here, Yale and Cambridge, went through significant organizational changes. Hence, another important body of literature we must consider is organizational change theory (DiMaggio and Powell 1983; Greenwood and Hinings 2006; Van de Ven 1986).

In this chapter, we start by considering external university factors—history and environment. We then move to internal factors—culture, policy, and organization—and consider the literature on organizational change. We conclude by presenting the theoretical approach of this book: there is no one formula for successful university technology commercialization. Universities affect their local economies through different mechanisms. The choice of mechanisms and their economic impact are influenced by the location (history and environment) in which the university is situated.

## EXTERNAL FACTORS

The ability of a university to transfer its technology to industry is influenced by two external factors: legislation and the relationships between institutions and organizations. The legislative and economic arenas on both the national and regional levels are important external factors that affect a university's commercialization capability (Lawton Smith 2006; O'Shea et al. 2005; Rahm, Kirkland, and Bozeman 2000).

### Historical Factors: National Level

In the United States, the federal Bayh-Dole Act has influenced regions in general and university-industry relationships in particular (Mowery and Sampat 2001b). The Bayh-Dole Act stipulates that the university owns the intellectual property rights for inventions that were based on a federal research grant. In Europe, each country created its own legislative incentives that formed a climate for university technology transfer. Inspired by the changes in the United States and the success of a few regions such as Silicon Valley, many European countries have attempted to implement similar policies in their regions (Lawton Smith 2006).

In 2004, the United Kingdom created its own incentives for technological and business development. For example, a small or medium-size enterprise can claim 225 percent corporation tax relief, and large corporations can claim 125 percent tax relief on R&D. Furthermore, the United Kingdom created the Technology Programme, which provides £320 million ($489 million) in the form of grants to R&D in degenerative medicine, biopharmaceutical processing, and bio-based industrial products, as well as the Bio-Wise program, a £14.5 million ($22 million), five-year Department of Trade and Industry program aimed at motivating manufacturing companies to increase their use of biotechnology (Department of Trade and Industry 2006; HM Revenue and Customs 2013).[1]

Although different countries and regions provide different support mechanisms for entrepreneurship and technology transfer, we find successful cases of university commercialization all across the globe. This indicates that policies are not the only factors that promote successful technology transfer and commercialization.

While in the United States, university-industry relationships have been viewed favorably and supported by the federal government since the inception of the states, the UK government did not view university-industry relationships favorably and hence did not provide any financial or legislative support until the 1990s. A full explanation of the historical and legislative factors in each of the case study countries is given in Chapters 4 and 5.

### *Historical Factors: Regional Level*

Regional-level legislation and economic policies also affect technology commercialization. In particular, the existence of a developed industry in a region can benefit universities' research. Industry collaboration comes in the form of student internships, future employment options for students, sponsored research, and potential invention licensing. This is particularly important in the United States, because states have the power to change local economic incentives, such as business taxes, and thereby affect the local industrial composition. Some states, such as California, Massachusetts, and North Carolina, have specific tax incentives for the high-tech industry. Some of this support can be seen in state funding of educational programs in schools and universities that both promote awareness of the industry and serve as the basis for a future educated labor force. Some examples of state-directed policies for the biotechnology industry are tax incentives that differ according to state. Massachusetts has a broad list of incentives in general and particular incentives for the life science industry. For example, it provides a 10 percent investment tax credit and a refundable research tax credit. California provides tax incentives of 15 percent (in-house) and 24 percent (outsourced), as well as R&D tax credits, with a 100 percent loss carried forward, a 7 percent job-creation tax credit, and a 6 percent manufacturing credit. Connecticut provides the R&D expenditures tax credit, which is 6 percent for companies with less than $100 million in income on R&D expenses incurred in Connecticut. In addition, Connecticut provides tax credit for firms' investment in human capital, machinery and equipment, and hardware and software. However, these incentives are general and not specific to life sciences (Mass Biotechnology Council and Boston Consulting Group 2002; Breznitz 2007; Mass Biotechnology Cluster 2013; Connecticut Economic Development 2013).

Regional-directed policies can also support innovation. The State of California collaborates with industry and state universities to develop jointly funded research programs. The State of North Carolina funded the North Carolina Biotechnology Center. The State of Connecticut created Connecticut Innovations, which invests in local companies to enhance economic development through its Seed Investment Fund and the

BioScience Facilities Fund.[2] The Seed Investment Fund provides seed capital to young Connecticut companies, and the BioScience Facilities Fund is a unique fund for the development of laboratory space. In the United Kingdom, such programs are available only at the national level.

In summary, while countries and regions differ in their incentives for universities' technology transfer, there are successful cases of university commercialization all over the globe. Therefore, university technology commercialization incentivizing policies are not the only factors that promote successful technology transfer and commercialization. Chapter 3 provides a historical and a political overview to explain the policies and economic frameworks within which Yale University and the University of Cambridge operate on both the regional and the national levels.

### Environmental Factors

Technology transfer from universities shapes innovation and economic growth, and it can be analyzed through, and affected by, institutional system-level theories. Environmental factors refer to the institutional system in which the universities operate. Here we focus on the different players within innovation systems.

Technology and information transfer in national and regional systems influences the rate of innovation (Lundvall et al. 2002; Nelson 1993; Freeman 1995, 2002). According to innovation systems theories, the environment in which universities operate—the relationships between nonfirm institutions and organizations in the region, such as government, trade associations, universities, and research institutes—influences their ability to innovate (Etzkowitz 1995; Nelson 1993). Note, however, that the regional system of innovation theory does not explore technology transfer specifically. Although universities have a central place in this theory, the theory does not examine the level of investment in, and the organization of, university technology transfer.

Moreover, universities, as part of a system of innovation, do not work in a void; they are influenced by the networks and relationships in their specific locality (Freeman 1995). These are symbiotic relationships, in which technology transfer influences innovation, and relationships in the system of innovation influence the ability to transfer technology. Hence, if we do not understand technology transfer, we cannot fully understand how systems of innovation work.

Cooke (2002, 136) insists on a correlation between the national and regional levels, too:

> Clearly, by no means all innovation interaction can or even should occur locally, but the rise of the entrepreneurial university and promotion of the so called "triple-helix" of interaction between industry, government and universities as a key feature of the knowledge economy testifies to the practical evolution of interactive innovation process.

Cooke and Morgan (1998) claim that some regions with innovative organizations, connected through joint research programs, policies, and social networks in an institutional milieu, "combine learning with upstream and downstream innovation capability" (71), thus making them *regional innovation systems*. This represents a regionally based innovative network of universities, colleges, and research institutes. In these regions, companies are able to access and test knowledge more easily. In regional systems of innovation, knowledge becomes the most strategic resource and learning the most important process of economic development (Lundvall 1994). To create economic development, high-level innovation and production processes need to be maintained in the region at all times. These processes are achieved through the constant learning and training of employees, an internal firm learning process that spills over to regional learning. The learning process occurs through local institutions such as universities.

According to innovations systems theory, some regions possess a particular infrastructure that allows them to achieve maximum regional learning. The learning and knowledge creation process is accomplished through a set of institutions that promote knowledge creation and learning by the local firms. There is a base of trust and understanding among firms, institutions, and individuals that differs from region to region and allows some regions to perform in a way that promotes their economic development.

Cooke and Morgan (1998) claim that successful regions consist of the following institutions: universities, basic and applied research laboratories, technology-transfer agencies, regional public and private associations, training organizations, banks, venture capital firms, and interactive small and large firms. They claim that the variety and level of research, as well as the network and communications among the different institutions and their cooperation, have a direct impact on the economic success of a specific locality (Cooke and Morgan 1998; see also Carlsson et al. 2002).

It is vital to understand that innovation results from the combined work of both public and private institutions. The public sector provides support to the private sector by enhancing production and distribution of technology and by reducing transaction costs (Lundvall et al. 2002). Consequently, universities and research institutes play an important part in the national and regional innovation process. This point can be effectively analyzed through Etzkowitz's (1995) "triple helix" model, which, with its focus on the communication networks among university, industry, and government, provides an argument for commercializing scientific knowledge: "As knowledge becomes an increasingly important part of innovation, the university as a knowledge producing and disseminating institution plays a larger role in industrial innovation" (Etzkowitz et al. 2000, 314).

Etzkowitz argues that universities, industries, and governments increasingly find themselves working together, as they understand that economic development can be achieved by creating and fostering innovative environments. Developing an innovative environment on a national or regional level can be achieved through the incorporation of university spin-outs, specific policies, networking among firms and government laboratories, and basic research conducted at universities. With globalization, corporations have increasingly discovered the advantages of tapping into the best research and practices in many places around the world. In other words, knowledge and top-quality research have geographical characteristics (Etzkowitz and Leydesdorff 2000). As such, they concentrate in specific locations and are based on regional learning, networking, and technology transfer. Not all nations or regions possess the best knowledge, and although globalization allows nations to import and export this knowledge, corporations still need to tap into specific localities. Therefore, Microsoft does not work with the UK government in general but rather with the University of Cambridge in particular. Similarly, even though Pfizer had its own R&D facility in Connecticut, the company chose to build a laboratory at Yale and to open another facility near Massachusetts Institute of Technology. Hence, the role of the university is greater than that emphasized by the triple helix model: "The markets and networks select upon technological feasibilities, whereas the options for technological development can also be specified in terms of market forces. Governments can intervene by helping create a new market or otherwise changing the rules of the game" (Etzkowitz and Leydesdorff 2000, 114).

In highlighting these relationships, this model treats the university's contribution positively, allowing little variation in the way different universities function. Most important, in this model, the university is differentiated only by whether or not it is active in the university-industry-government networks.

Universities do not operate in a vacuum, however; they are accepted as part of the system of innovation and as one of the channels for knowledge transfer. Though both the system of innovation and the triple-helix theory acknowledge the importance of the environment in which universities operate, they do not view the university as the main actor that can affect regional economic success. It is evident from these theories that universities operate in a specific environment, which is one of the factors influencing their ability to transfer technology. In both our cases we find the local university to be a strong actor in the local system of innovation. The University of Cambridge is a well-known university, but, as we will see, it has a particular impact on the United Kingdom in general and on the east of England in particular. Similarly, Yale University in Connecticut plays an important part in its state's and region's decision-making processes.

## INTERNAL FACTORS

Three important internal factors influence the technology-transfer capability of a university (Bercovitz et al. 2001; Clark 1998; Etzkowitz 1998; Link and Scott 2005; O'Shea et al. 2005; Schoenberger 1997; Shane 2004). See Figure 2.1.

The first is the university's entrepreneurial culture, which is shaped to support risk taking, innovation, new business creation, and a willingness to collaborate with industry (Bercovitz and Feldman 2007; Clark 1998; Etzkowitz 1998; James 2005; Kenney and Goe 2004; Schoenberger 1997). The second is technology-transfer polices, which affect the university's ability to patent, license, and spin out companies (Blumenthal et al. 1996; Lawton Smith 2006; Link and Siegel 2005; Shane 2002, 2004; Siegal and Phan 2005; Thursby and Thursby 2005; Zucker, Darby, and Armstrong 2002). The third factor is the university's technology-transfer office (Bercovitz et al. 2001; Clarysse et al. 2005; Link and Siegel 2005; O'Shea et al. 2005; Owen-Smith and Powell 2004; Siegel et al. 2004), viewed through organizational change theories (DiMaggio 1988; Greenwood and Hinings 2006; Meyer and Rowan 1977; Pfeffer 1978; Scott 2003; Scott and Davis 2007; Tolbert and Zucker 1996).

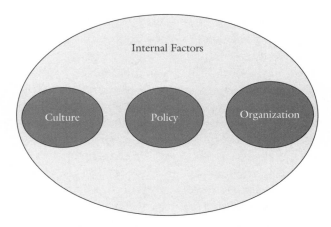

FIGURE 2.1 Internal university technology commercialization factors
NOTE: Three important internal factors influence the technology-transfer capability of a university.

### University Culture

Like *technology transfer*, the term *culture* is often ill defined in the literature, leading to confusion. When referring to culture, some scholars focus on the university's leadership (Schoenberger 1997; Kenney and Goe 2004), some on the culture of the institution itself (Clark 1998), and others on the general culture in the region in which the university operates (James 2005). What is apparent from the literature is that universities' success in transferring technologies to the public domain is clearly based on the university environment and entrepreneurial culture. When individuals study or work at the same university, a common background is created, which leads to trust and allows for an exchange of information. For example, many university-industry relationships are based on the involvement of faculty members in capital formation projects, such as industrial consulting and new firm formation.

Etzkowitz (1998) suggests that the role of the university as entrepreneur produced a new breed of academics, the entrepreneurial scientist who can pursue truth and profit at the same time. To support the change from an ivory tower to an entrepreneurial university, many scientists justify the relationship with industry as a financial resource that will continue to support academic research.

Studies have found that entrepreneurial scientists, as highlighted by Etzkowitz (1998), have specific characteristics. Roberts (1991), who has focused on star scientists, found that MIT academics, who are considered by many to be the most entrepreneurial, have a high desire for independence and a need to achieve. Zucker, Darby, and Peng (1998) found that academics who collaborate with industry have more citations than academics who do not, which emphasizes the importance of university-industry relationships where the advantages of these relationships work for both sides. If you are a well-known academic but also work with industry, your chances of recognition are higher, and vice versa, so industry will be interested in your work.

Clark (1998) emphasizes the cultural climate of the entire university— all departments and research centers, not just the leadership—as key for entrepreneurial activity. He claims that, "particularly for universities, we stress *interactive instrumentalism*" (145). Referring to interactive instrumentalism, Clark emphasizes the interactions of all elements and policies, which contribute to the transformation of the university: "Transformation requires a structured change capability and development of an overall internal climate respective to change" (145).

Extending the importance of the particular university culture to the success of university-industry relationships and regional growth, James (2005) claims that it is not enough for firms and institutions to promote collaboration. To maintain growth and change, a new behavior of universities is required. According to James, corporate or institutional culture, which helps stabilize organizations, is part of a regional cultural frame. Thus, a regional culture includes "(1) the individual corporate cultures; (2) a regional industrial culture; and (3) the broader regional culture in which these are set" (James 2005, 1199).

Studies show that the culture within one's department affects technology commercialization. Kenney and Goe (2004) found that a scientific department's culture as well as its individual colleagues influence the scientists' entrepreneurial spirit. Moreover, the university itself influences the university's "professional entrepreneurship and corporate involvement" (Kenney and Goe 2004, 704). Furthermore, Bercovitz and Feldman (2005) found that the culture in which the scientist trains, works, and collaborates has a direct influence on the scientist's ability to innovate and commercialize technologies. In particular, they suggest that an academic whose departmental chair and colleagues file inventions, thus creating an entrepreneurship culture, is more likely to file inventions and participate in technology-transfer activities. Hence, many universities see changes in commercialization not only through policy changes; such changes come when university members view working on applied research and commercializing technology as acceptable and desirable. At Yale, technology-transfer office (TTO) employees and department chairs engaged in discussions with faculty to explain new technology-transfer processes, but they also made sure to reach out to researchers who had exhibited previous prejudice toward commercialization.

Corporate culture has received the most attention in organizational studies and business literature, but many insights may be tangible to the institution of the university. In management theories, corporate culture is characterized as the behavioral norms that affect social interactions in the firm. In her book *The Cultural Crisis of the Firm*, Schoenberger (1997) states that an organization's cultural base affects the way in which the organization makes decisions about issues such as strategy, outlook, and cooperation with other players in the local economy. Accordingly, Schoenberger argues that "the interpretive process is deeply cultural" (204). Kunda (1992) argues that corporations use culture to create and promote a specific work environment. Accordingly, corporations create specific environments, encouraging individual creativity and commitment to the firm. Schoenberger also argues that culture plays an important role in the organizational change process and is led by powerful decision makers. Decision makers have central positions in the organization and can make life-altering decisions that will affect the future of the organization. However, the process of decision making is based on social relations and a consciousness of firms' decision makers (Schoenberger 1997).

In summary, theories of cultural change explain how an organization's strategy is influenced by the commonalities or norms of individuals in the organization (see Figure 2.2). These studies discuss the importance of the culture of the organization itself as well as the collegiate, departmental, university, and regional entrepreneurship environment (Kunda 1992; Schoenberger 1997).

Other studies analyzing university culture, such as Clark (1998), Etzkowitz (1998), Bercovitz and colleagues (2001), and Kenney and Goe (2004), emphasize the importance of an entrepreneurial culture as the basis of university technology-transfer success. Hence, for successful technology commercialization, the individual's view on commercialization, the way the university administration views and organizes technology transfer, and the way in which the region accepts and responds to university commercialization are all important factors. Creating an entrepreneurial culture affects the university and its researchers in multiple ways: it allows faculty to work on applied research and to accept the ability of academic

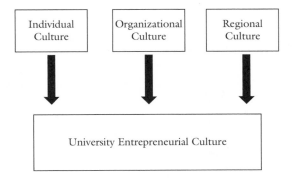

FIGURE 2.2 Technology-transfer culture

NOTE: An organization's strategy is influenced by the commonalities or norms of individuals in the organization.

research to make profit and a public impact, and it creates a possibility to found a new company based on university research.

### University Policy

The second internal factor examines the importance of academic policies that affect university-industry relationships. In particular, for technology transfer, these include regulation of intellectual property rights related to patenting, licensing, and spinning out companies.

Intellectual property rights (IPR) policy affects the university technology-transfer, commercialization, and economic development functions. At universities, this refers to copyright of academic publications—that is, journals and books, and patents filed by the university for an invention that was the result of university research. Although definitions of IPR differ among universities given each institution's history, culture, and technology-transfer organization, one way to view IPR policies is through inventions ownership. At one extreme, all the universities' inventions are owned by the universities; at the other extreme, ownership depends on the funding source for the research project (Siegal and Phan 2005).

Why is ownership important? Because the owner of the technology can license or develop that technology. The ability of the owner to develop or license out a technology is significant, especially when referring to technology that was developed using public funding. In a time when universities are urged to become involved and have economic impact, the decision to further develop a technology is highly important. Both development and licensing require resources; particularly for universities, the ability to communicate and negotiate with corporations, funding to patent technology, and even the basic understanding of the technology quality, is crucial. If a university owns the technology but has no resources or basic business understanding of how to license the technology, the chance of commercialization is slim to none. Hence, understanding what policy choices Cambridge and Yale made in regard to IPR ownership is vital in our attempt to understand their success in contributing to local economic development.

A good example of differences in IPR policies is documented in Lawton Smith (2006) as well as in Shane (2004). According to these studies, in Sweden and Italy, faculty and staff own their IPR, whereas in the United Kingdom each institution has its own rules with regard to intellectual property (IP) ownership. As mentioned previously, individual ownership of IP can be problematic. In Sweden, individuals own their IPR, and hence, they are responsible for the patents' costs, which results in less patenting and commercialization (Shane 2004). In Italy, although individual researchers own the IPR of their inventions, most universities lack the managerial experience and the commercial attitude to assist them with the commercialization process (Geuna and Nesta 2006; Lawton Smith 2006).

Royalties are the sum of money paid to the proprietor or licensor of IP rights for the benefits derived, or sought to be derived, by the user (the licensee) through the exercise of such rights. In the case of universities, royalties refer to the payments that the university and inventors receive from the licensing or equity of a university patent. When commercializing a university invention through the technology-transfer office, universities can either license the technology to an existing company or form a new company based on the invention.

On the basis of fifty-five interviews with ninety-eight individuals, including managers, administrators, and scientists at five major research universities in the United States, Link and Siegel (2005) found that a higher share of royalties enhances technology licensing. According to Link and Siegel, changes in faculty incentives change their behavior. By allocating a higher share of royalties to the inventor (the faculty member), universities will license more inventions to existing companies. The work of Di Gregorio and Shane (2003) reinforces Link and Siegel's point about the correlation of royalties and licenses. Those authors found that there is an inverse correlation between the percentage of an inventor's royalties and the number of university spin-outs each year. Furthermore, Shane (2004) claims that allocating a lower share of royalties to inventors promotes spin-out capability. Thus, an allocation of fewer royalties to inventors will reduce the number of licenses and promote the creation of spin-outs. According to Shane, the reason for the inverse relationships in royalty share and spin-out formation is based on the allocation of royalties at the university. The higher the inventor's royalty share, the higher are the opportunity costs related to spinning off a company, and therefore the lower are the incentives to do so.

Examining the reasons behind the licensing activities of faculty and researchers, Link and Siegel (2005) found that full and associate professors, being already tenured in most universities, do most of the invention disclosures and patenting. Accordingly, Link and Siegel claim that the creation of policies that will encourage and protect junior faculty will increase universities' patents and licenses.

When we examine both the University of Cambridge and Yale University in light of these theories, we find that indeed Cambridge provides a higher percentage of royalties to inventors, thus promoting licenses over spin-outs, whereas Yale provides a lower share of royalties, which promotes the creation of companies.

Di Gregorio and Shane (2003) also identified another policy influence: those universities that are willing to take equity for patenting and licensing expenses have a start-up rate that is higher than those universities that refused equity. In his book *Academic Entrepreneurship*, Shane (2004) summarizes university policies and their influence on university spin-outs. In addition to the reasons mentioned already, he stresses the importance of universities allowing exclusive licensing, offering leaves of absence, permitting the use of university resources, allocating a lower share of royalties to inventors, and providing access to pre-seed-stage capital.

Another aspect of technology-transfer policy that has impact on a university's spin-out capability is the extent to which research collaboration is permitted. Faculty collaboration with industry through consulting or research projects affects industry-sponsored research as well as the university's culture and view of applied research. University-industry research collaboration promotes financial support from industry, which supports students or provides grants for particular research. Moreover, if the research results in an invention, industry will purchase the invention equity or license the technology from the university (Blumenthal et al. 1996; Thursby and Thursby 2005; Zucker, Darby, and Armstrong 2002).

Shane (2004) adds that the proportion of industry's contribution to research funding is a predictor of the level of university spin-outs.

Spin-out formation grows with the proportion of industry funding. Though we do not possess information on the exact share of industry funding for each university, we can see that between 1999 and 2010 at both Cambridge and Yale, the share of grants and contracts of the total income was around 30 percent, which, according to Shane, means that both should have a similar number of spin-out companies. Interestingly, Yale's share of grants and contracts went down, from 28 percent to 21 percent, and Cambridge's stayed the same, around 35 percent. Between 1999 and 2010 Cambridge's income grew by 138 percent, including an increase of 149 percent in income from grants and contracts, whereas Yale's income grew by 156 percent. Even after the growth, Cambridge's income was only 41 percent of Yale's. See Figure 2.3.

The fact that income from and collaboration with industry have an impact on the number of university spin-outs makes Yale and Cambridge interesting cases. The data from both institutions support Shane's (2004) findings. Cambridge's income is lower than Yale's, and the university produced sixty biotechnology spin-out companies by 2012, compared with

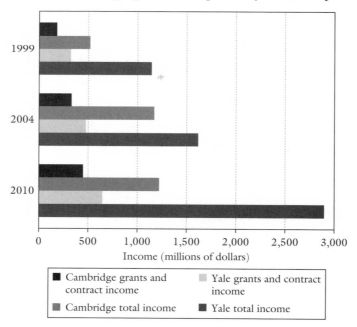

FIGURE 2.3  Cambridge and Yale income changes, 1999–2010

SOURCE:  Office of Institutional Research (2001); University of Cambridge (2004); Yale University (2011–2012); University of Cambridge (2013).

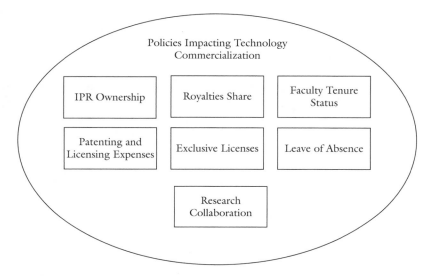

FIGURE 2.4 Technology-transfer policy
NOTE: There are many policies that impact university commercialization.

seventy-one for Yale. A possible reason for the differences in university spin-out capability between Cambridge and Yale could be that Yale may have had a catch-up spurt, while Cambridge's lower rate of growth may reflect previous use of commercialization potential. That might have been true if this trend did not continue. An analysis of the universities' spin-out pace through 2012 shows that Cambridge never went back to form more than one or two spin-outs a year, and firms' value (measured by investment) is smaller compared with that of Yale (see Chapter 4).

As we can see from Figure 2.4, there are many policies that affect university commercialization. What we don't know is to what extent these policies are needed, and how these policies should be composed, for a positive impact on university technology commercialization.

### *University Technology-Transfer Organization*

The third internal factor in the output of a university technology transfer is its institutional organization and management. In particular, the university's technology-transfer office is the university resource with the strongest impact on the creation of spin-out companies. Studies on organizational change as well as studies on university technology-transfer offices are relevant in understanding technology transfer at universities. Numerous studies have found that the characteristics of the university technology-transfer office influence the level of spin-out activity for the institution, particularly the size, income, experience, and past success (Owen-Smith and Powell 2001; Bercovitz et al. 2001; Carlsson and Fridh 2002; Chapple et al. 2005; Foltz, Barham, and Kim 2000; Lockett and Wright 2005; O'Shea et al. 2005; Shane 2004; Siegel, Waldman, and Link 2003a).

## ORGANIZATIONAL CHANGE

Technology-transfer offices are organizations within institutions (universities). This section examines the literature of organizational change. In particular, we focus on reasons for change and the impact of individual behavior on the organizational change (Greenwood and Hinings 2006; Van de Ven 1986). Studies of organizations became commonplace during the 1970s (Greenwood and Hinings 2006; Pfeffer 1978; Scott 2003; Scott and Davis 2007; Van de Ven 1986). It was only in the mid-1980s that organizational change became a significant topic of its own in the organizational studies literature.

Previous studies have determined several causes of organizational change (Bercovitz and Feldman 2008; Dacin, Goodstein, and Scott 2002; DiMaggio and Powell 1983; DiMaggio 1988; Greenwood and Hinings 2006; Pfeffer 1978; Tolbert and Zucker 1996; Tushman and Romanelli 1985; Van de Ven 1986; Whelan-Berry, Gordon, and Haining 2003; Bandura 1977; Meyer and Rowan 1977; Schein 1985; Scott 2003; Scott and Davis 2007). First, individuals within the organization effect change. The imprints individuals carry from their historical and environmental backgrounds, as well as their willingness to adopt new norms and change, are vital to the process of organizational change (DiMaggio and Powell 1983; Greenwood and Hinings 2006; Schein 1985; Whelan-Berry, Gordon, and Haining 2003; Dacin, Goodstein, and Scott 2002; Bandura 1977). Every organization has a flow of people into and out of it. Incoming people bring new ideas and norms, and the exit of employees allows those new ideas to set.

In their study of university faculty members' engagement in technology commercialization, Bercovitz and Feldman (2008) found that the workplace environment affects the individual's behavior in favor of or against commercialization and entrepreneurship. The behavior of the department chairs and co-researchers in a department has a direct impact on the researchers' participation in technology commercialization. The notion that individuals are part of a larger process of change is not limited to organizational studies. Studies on social networks and systems of innovation have found knowledge transfer to be vital for economic development, which is based on the movement of entrepreneurs and employees between companies and institutions (Amin, Thrift, and ESF Programme on Regional and Urban Restructuring in Europe 1994; Boucher, Conway, and Meer 2003; Casper 2007; Freeman 1995; Murray 2002; Owen-Smith and Powell 2004; Owen-Smith et al. 2002; Nelson 1993). Technology commercialization organizations at universities are not different. Both Yale and Cambridge have gone through organizational change. Moreover, similar to many other higher education institutions, these two universities have a constant inflow and outflow of faculty and staff, and the individuals who research and teach at these universities, as well as the university organizations that control technology commercialization, have a direct impact on the ability of the institutions to bring technology to market.

According to Owen-Smith and Powell (2004), the adaptation of the new organizational form depends on management's susceptibility and connections to external networks. Thus, the individual connections of the organization's leadership can help with and support the change. Studies have shown that the ability of a company to succeed is based on a limited number of people in the region's social network. Studies by Myint and Vyakarnam have shown that at Cambridge, there is a tight network of individuals; in many cases the success of a company within the region is based on connection to three or four individuals (Myint, Vyakarnam, and New 2005). At Yale, individuals that made an impact were the president Richard Levin, Greg Gardiner (and later Jon Soderstrom) from the technology-transfer office, as well as Bruce Alexander, who was brought in to manage the Office of New Haven and State Affairs in 1998.[3] Levin, an economist who studied science and technology, was highly versed in the role of universities in economic development:

> The Stanford Paradigm was part of what I knew, off course it was influential. But also, I studied the economics of science and technology, one of the positions I resigned when I became president was the editor of *Research Policy* [the journal]. I engaged and knew all the literature about, at the time. I knew all the literature on route 128 and Silicon Valley and their emergence and the European hubs. Hence, I was knowledgeable and influenced by what I knew. (Interview with Richard C. Levin, 2013)

Moreover, hiring just the right person to head the Office of Cooperative Research was an important ingredient in the ability of the university to change its commercialization practice. The person whom President Levin and Provost Richard chose was Gregory Gardiner, a former faculty member who had left Yale for a position at Pfizer. Gardiner had seen varying models of university-industry relationships and was a strong advocate for applicable research.

Another impetus for change in an organization is an acknowledgment of organizational failure. Studies show that for organizational change to occur, however, decision makers must be convinced of positive results, which will justify new forms of organization and resource allocation (Tolbert and Zucker 1996; Greenwood and Hinings 2006). In the case of Yale University, the murder of a student on campus in 1998 forced the institution to take a grim look at its relationship with the city of New Haven (Atlas 1996; Ball 1999; Blumenstyk 1990). Although New Haven is very small, the city shared the problems of many larger cities, such as higher crime rates, unemployment, and urban development issues. Yale's response was to enact a four-step economic development program, and the change in technology transfer played a large part in this program. It had become clear to university leaders that for Yale to become what Yale's president at the time, Richard Levin, called "a contributing citizen" would benefit both the university and the local community (Levin 1993, 2003).[4]

Organizational change can also occur as a result of market or institutional pressure (DiMaggio 1988). At Cambridge, change came directly from the British government in the form of reports and funding for university-industry relationships. But why do different organizations have different capabilities to adapt to organizational change (Greenwood and Hinings 2006)? One reason is that organizations are set in their ways, and change may require force or pressure (Meyer and Rowan 1977). At both Cambridge and Yale, continuous requests from department chairs and deans, as well as funding, contributed to the process of change.

Following the Dearing Report of 1997 and the Sainsbury Report of 1999—both of which dealt with research and development and its impact on the economy—the University of Cambridge started its technology-transfer organizational changes by creating different offices to deal with technology commercialization and then centralizing them into one office: the Research Services Division. Changes continued with the Higher Education Innovation Fund of 2001, which allowed the university to decentralize its technology commercialization activities. Other changes followed and continued until 2006.[5] Hence, Cambridge changes were a direct response to certain initiatives, thus resulting in a punctuated change that took about ten years, whereas Yale's organization change took place in just three years, from 1993 to 1996. The time frame and the velocity of changes matter, and these issues are discussed later in this book.

## TECHNOLOGY-TRANSFER OFFICE

The technology-transfer offices in universities have four main purposes: (1) to evaluate inventions and determine whether they are patentable; (2) to patent the inventions; (3) to license the technology; and (4) in some cases, to assist in the creation of spin-out companies. The technology-transfer office's responsibilities are quite loose and open to interpretation, however, and they differ significantly between universities. Some universities will patent only a technology that has market demand to which it can be licensed. For many, the spin-out of companies is not a priority; the goal is to garner income from licensing their patents. Furthermore, in many cases the professionalism and actions of the technology-transfer office affect the likelihood of being able to license a technology.

The level of resources associated with the technology-transfer office affects its commercialization ability. Several studies showed that technology-transfer offices that have personnel with higher levels of education and business experience tend to have better understanding of the technology and negotiation processes with firms. Understanding business and product development allows for more flexibility and trust and promotes the willingness of inventors and investors to work with that office (Lockett and Wright 2005; Shane 2004; O'Shea et al. 2005). Since university and industry have different business perspectives, highly educated employees in the technology-transfer office who have knowledge of both technical and business jargon reassure both inventors and investors that their product is getting the best available treatment.

Moreover, the professionalism of the TTO affects faculty disclosure rates and commercialization interests. A study by Owen-Smith and Powell (2001) showed that larger and more experienced TTOs are able to provide personal and professional care when working on faculty inventions, thus encouraging faculty to disclose and patent technologies. By 2004, Yale University's technology-transfer office employed eighteen people, each with five to seven years of industry experience. At the same time,

Cambridge Enterprise (Cambridge's TTO) had eighteen employees, fifteen of whom had no industry experience. The differences in the two offices in terms of employee experience surfaced in my interviews, which revealed constant complaints about Cambridge Enterprise's lack of business understanding.[6] Hence, the business experience and knowledge base of TTO employees have more weight than the sheer number of TTO employees. What is known today as "The Silicon Fen" is based on many technologies that came out of the University of Cambridge's TTO, which for many years operated with only two employees.

Both Clarysse and colleagues (2005) and Lockett and Wright (2005) found that the business development capabilities of the technology-transfer office positively influence start-up formation. The variables they found to be of the greatest importance were the marketing, technological, and negotiating skills of the technology-transfer office staff, the establishment of a clear administrative process for spin-out companies, a clear due diligence process, and availability of competent staff to administer these processes (Lockett and Wright 2005; Clarysse et al. 2005).[7]

Another factor that relates to the availability of resources is the use of outside lawyers. Siegel, Link, and Waldman (2003a) found that spending more on outside lawyers reduces the number of licensing agreements but increases revenues. The authors hypothesized that hiring external lawyers allows TTO staff to spend more time on connecting the invention to the right firm, which results in a successful license and higher revenues. In examining the TTO staff, the authors noted that when TTO officers received incentive-based compensations, there was higher licensing activity. Moreover, the quality of staff, specifically their business and marketing skills, affects the university's technology-transfer productivity. The fact that many TTO staff lack business and marketing experience reduces the university's technology-transfer productivity (Siegel, Waldman, and Link 2003a; O'Shea et al. 2005; Owen-Smith and Powell 2001; Shane 2004).

Importantly, technology-transfer offices, much like universities themselves, differ in their compensation abilities. For example, technology-transfer offices vary in their ability to offer wages and benefit packages to attract high-level employees, who are highly compensated in the private sector. Accordingly, many technology-transfer offices differ by the education and experience of their employees. In many, industry experience and a doctorate in the sciences is a necessity. To solve the issues of staff compensation, some public universities' technology-transfer offices became private organizations that are fully owned by the university. For example, Isis at the University of Oxford, United Kingdom, and Yissum at the Hebrew University, Israel, are private organizations. On the basis of interviews at Cambridge and Yale, we know that Yale, as a private university, could offer competitive salaries to its TTO employees, whereas Cambridge Enterprise, which until 2006 was a not an independent entity, could not.

The size as well as the age of the technology-transfer office has also been found to affect university technology transfer (Carlsson and Fridh 2002, 230). According to Chapple and colleagues (2005), UK technology-transfer offices show low levels of absolute efficiency. Their study found that older offices appear to be less efficient by the number of licenses compared with research income and invention disclosure. This may suggest that older TTOs have not changed or adapted to the "third role" of universities. Furthermore, their study shows that larger TTOs are less efficient. Therefore, larger universities are more general and include many fields of technologies, and we know from other studies (Owen Smith and Powell 2001) that technology transfer in the life sciences is very different from the physical sciences. This finding suggests the possible need to invest in smaller, specialized offices instead of the general growth of the office (Chapple et al. 2005; Owen-Smith and Powell 2001).

A study by O'Shea and colleagues (2005) found that the historical background and past technology-transfer success of each university is related to future capabilities and options for the university with regard to spin-out capability. When a technology-transfer office has successfully seen an invention go through the commercialization process, and sees returns in the form of royalties, the office is strengthened and motivated to continue with the commercialization process. Yale University has seen success in technology commercialization via its patenting of Zerit, one of the drugs used in the treatment of HIV/AIDS patients. However, the University of Cambridge, which spun out a similar number of biotechnology companies to Yale, and has been engaged with industry since the nineteenth century, has not had a large revenue-making patent in biotechnology such as Zerit.[8]

In summary, technology-transfer organizational theories discuss the specific characteristics of technology-transfer offices, which allow for better output in the form of licenses and spin-outs. These studies focus on offices' resources and personnel (see Figure 2.5). A summary of existing studies finds that the TTO's budget determines the ability of the office to hire qualified employees. In particular, studies showed that employees with PhD's in sciences and business experience are better able to evaluate technologies and negotiate with local businesses. Moreover, the resources

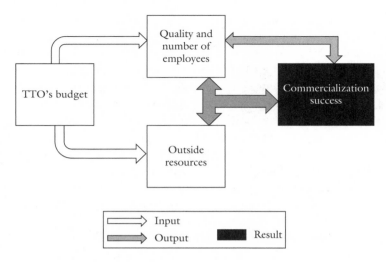

**FIGURE 2.5**  Technology-transfer organization
NOTE:  The TTO's resources impact its ability to commercialize technology.

of the office provide a potential larger number of employees, which some studies show may impact technology commercialization. Moreover, financial capabilities allow offices to use outside resources like outside lawyers in general or for complex cases. Last, studies show that a success in the form of revenues from licenses or equity provides an office the opportunity to learn what the possible gains are.

## THEORETICAL APPROACH

A review of existing literature on the factors that affect universities' ability to commercialize knowledge reveals that there is a gap that existing theories do not sufficiently explore. In particular, (1) studies focus on individual university cases which are hard to generalize from; (2) data on which studies rely on varies, and studies' definitions of technology commercialization are different; (3) change in technology transfer is viewed positively with an approach of "one formula fits all"; (4) many of the studies do not make the connection between technology commercialization and economic development; (5) many studies do not view these factors with a general analysis that includes history, environment, and internal university factors. Hence, different studies on different universities have created a set of contradictory results. A review by Bagchi-Sen and Lawton

Smith (2012) focuses specifically on the difficulties in generalizing case studies of universities' commercialization and economic development as well as the problems with sources and data.

The focus on individual cases specifically created an analysis of "success" cases for university technology commercialization. By selectively focusing on positive success stories and failing to address less successful cases, these studies may distort our view of the technology-transfer process, thus leading us to misunderstand the casual relationships involved and to emulate the wrong model. Different case studies use different data sources and different definitions of technology commercialization, which makes generalization of the phenomenon impossible. The data are usually based on case studies, which rarely use similar definitions for even something as basic as university spin-out. Existing theories argue that technology commercialization changes to policies and processes have a positive impact on the university itself as well as on university-industry relationships. Hence, if university X improves technology commercialization, we can use that method at university Y to bring technology to the market. Importantly, given different histories, economic settings, and players, not all the factors that worked for one university will necessarily work for another. Adopting similar policies from a successful case of technology transfer does not guarantee similar results. Hence, the existing literature does not adequately explain how universities' technology commercialization investments and organization affect their ability to disseminate academic ideas to the private market.

Moreover, as Lawton Smith (2006) concludes in her book *Universities, Innovation, and the Economy*, universities today face unrealistic expectations in regard to economic development and public contribution. Lawton Smith rightly concludes that universities are expected to provide solutions to economic growth problems. However, few existing studies consider universities' capabilities or the possibility that their actions may have neutral or negative effects (Bagchi-Sen and Lawton Smith 2012; Lane and Johnstone 2012; Rothaermel, Agung, and Jiang 2007). The following chapters of this book fill this gap by focusing on universities' ability to transfer knowledge and commercialize inventions in order to explore variation among universities.

The theoretical approach of this research assumes that universities can influence their local economies, thus focusing on the differences in universities' ability to commercialize technologies. My main research question was, How do universities' organization of, and investment in, technology transfer affect their ability to disseminate academic ideas to the private

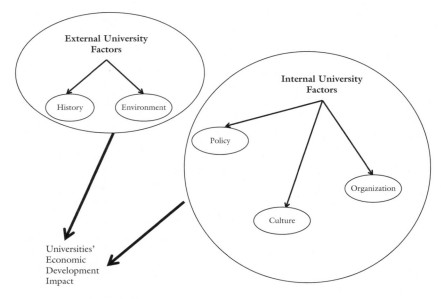

FIGURE 2.6 Theoretical approach

NOTE: The university analysis includes both internal and external factors.

market? While analyzing the university technology commercialization culture, policy, and organization within the wider frame of each university's regional history and environment, I found that each university has its own approach to investing and organizing knowledge transfer, which resulted in different mechanisms of commercialization. Moreover, I found that university knowledge transfer culture, policy, and organization can affect the university's ability to patent, license, spin out companies, and form relationships with industry. See Figure 2.6.

Providing a new view of university-industry relationships, my research began with an analysis of the region in which each university operates, followed by an analysis of university policies and processes in technology transfer and commercialization. The following chapters evaluate each of these categories on two different universities in different locations, examining university technology-transfer efforts through patenting, licensing, and spin-out companies and considering how each location affected the university's policies and processes. Finally, I discuss how these changes to technology transfer at universities shape the role of the university and the trajectory of its region.

## CHAPTER THREE

# The US and UK Technology Commercialization Framework

Universities do not operate in a vacuum. They are influenced by the environments in which they operate—and Cambridge and Yale exist in vastly different environments. The United States and the United Kingdom have historically taken different views on the role of the university, which has led to differences in the development of science and technology (S&T) policies in each country and has affected university research and the ability to commercialize technology. This chapter provides a general explanation of this history and considers the forces outside the two universities that affect their ability to make local economic development contributions.

In this chapter, we first consider each country individually, discussing its unique S&T policies and how these policies have affected the ability of local universities to commercialize technology. Then, we compare S&T policy and funding in the United States with those of the United Kingdom. In the United States, we find that higher educational institutions and research in general have been part of the country's production system since the 1880s, as reflected via funding and use of university research in government programs. In the United Kingdom, however, the potential commercial value of university research was not recognized until the 1950s, and its benefits for industry and commercialization were not fully realized until the 1990s. Thus, these two very different histories had a direct impact on the abilities of Yale and Cambridge to commercialize technologies and contribute to local economic development.

UNITED STATES

The view of research as a national commodity started in the United States in the late 1880s. Universities were seen as a way to increase agricultural and industrial productivity. During the two world wars, universities and basic research played a vital role in the political, social, and economic aspects of the country. Since World War II, policies and programs established over the years have helped shape the current S&T policy of the United States and have established the necessity of academic research.

### Science and Technology Policy

The US academic system is constructed from myriad colleges and universities across the country that vary in size, ownership, and specialization. Until the twentieth century, American universities had limited technology capabilities and were small, thus restricting any commercialization efforts (Shane 2004).

At the beginning of the twentieth century, science began to play a greater role in US industry. This was the result of two major events: the emergence of professional engineering schools in the 1800s and the building of a canal system during the same period. The formal use of university research for national needs began with the enactment of the land-grant colleges through the Morrill Acts of 1862 and 1890 (Rahm, Kirkland, and Bozeman 2000). The first Morrill Act provided grants in the form of federal land to each state for the establishment of a public institution to fulfill the act's provisions.[1] Through the provision of funding for the creation of universities that conduct applied research with the purpose of transferring the technology to the public, these land-grant colleges have become the cornerstone of the US model of enabling technology transfer.

By the late 1880s, although a number of institutions had begun to expand the traditional classical curricula, higher education was still widely unavailable to many agricultural and industrial workers. The Morrill Act was intended to reach some of those workers, providing a broad segment of the population with a kind of higher education that had direct relevance to their daily lives. The goal of the land-grant colleges was to teach agriculture, military tactics, and the mechanical arts, as well as classical studies, so that members of the working classes could obtain practical skills and knowledge.

The land-grant colleges were the first higher education institutions in the United States that saw technology transfer as a public good. With the extension service component requiring the dissemination of academic research, added through the Smith-Lever Act of 1914 (at the time mainly in agriculture), the land-grant colleges and universities generated a system in which the collaboration of three spheres (science, government, and industry) was highly acceptable.[2]

During World War I, universities were involved in the war effort, carrying out R&D projects sponsored by the government. During World War II, US scientists actively built new weapons and laid the groundwork for science-government relations. These collaborative efforts were described in the 1945 Vannevar Bush report "Science, The Endless Frontier," and the 1947 President's Science Research Board report "Science and Public Policy," on the contribution of science to the nation.

These reports led to the creation of the National Science Foundation (NSF) in 1950, "to promote the progress of science; to advance the national health, prosperity, and welfare; to secure the national defense" (NSF 2005). The establishment of the NSF marks one of the more important historical factors influencing university technology transfer in the United States. During both the Cold War and the Korean War, there were increases in military R&D government investments, and in 1958, the National Aeronautics and Space Administration (NASA) was formed through congressional legislation (Rahm, Kirkland, and Bozeman 2000). Some scholars argue that the federal government's spending on military R&D during that period was an important factor in the development of the high-tech and biotech industries in the United States (Markusen 1991). Over the years, the federal government has become the largest supporter of academic R&D, providing 63 percent of the total funding base (NSF 2013). The National Institutes of Health (NIH), within the Department of Health and Human Services of the federal government, is the largest funder of US universities' science and engineering programs.

Still, it was not until the 1980s that there was an institutional move to support a relationship between industry and universities. The decline in the US industrial growth rate in the 1970s and the strength of international competition created a debate over the government's investments in military R&D and over industry's ability to commercialize this research. Calls for stronger links between universities and industry to improve competitiveness resulted in a series of acts, created during the 1980s, to involve universities in applied research, with direct impact on the US industrial growth rate.

The Bayh-Dole Act of 1980 was considered the most influential in increasing technology transfer at universities. The act gave universities the rights to federally funded inventions. It states:

> It is the policy and objective of the Congress to use the patent system to promote the utilization of inventions arising from federally supported research or development; to encourage maximum participation of small business firms in federally supported research and development efforts; to promote collaboration between commercial concerns and non-profit organizations, including universities; to ensure that inventions made by non-profit organizations and small business firms are used in a manner to promote free competition and enterprise without unduly encumbering future research and discovery; to promote the commercialization and public availability of inventions made in the United States by United States industry and labor; to ensure that the Government obtains sufficient rights in federally supported inventions to meet the needs of the Government and protect the public against non-use or unreasonable use of inventions; and to minimize the costs of administering policies in this area. (Section 200, Title 35)

By affording universities the right to inventions resulting from federally funded research, the act encouraged them to commercialize their technology, thus leading to an increase in university patenting, licensing, and spin-out formations. According to Mowery and Sampat (2001b), the 1970s saw a similar increase in university patenting related to national incentives.

Following the Bayh-Dole Act, additional programs were formed to improve university-industry collaborations and to assist in funneling university research into the public sphere. As a result of the Small Business Innovation Development Act of 1982, the Small Business Research and Development Enhancement Act of 1992, and the Small Business Innovation Research Program Reauthorization Act of 2000, the technology program within the US Small Business Administration was created. The technology program is directed to the encouragement of research and development in small firms, through the Small Business Innovation Research (SBIR) program and the Small Business Technology Transfer (STTR) program.

Established in 1982, the SBIR program provides funding for small firms and start-ups to sponsor their research and development stages.[3] Hence, for many university-based inventions and spin-outs, SBIR is an important funding source. The program funds the transition from a research-based idea to a prototype that many industrial and venture capital companies find difficult to support (Bartik 1985; Audretsch, Weigand, and Weigand 2002). Companies eligible for the program are American owned and independently operated; they are for profit, meaning that the principal researcher is employed by the business; and the company has fewer than five hundred employees. Eleven federal agencies, including the NIH, the NSF, and NASA, set aside a portion of their extracurricular research and development budgets each year to fund research proposals from small science- and technology-based firms. The program consists of three phases. In phase one, a small business can receive up to $150,000 to establish a feasibility study of ideas that appear to have commercial potential. Phase two provides a contract for up to $1,000,000 to continue research in the same technical area. The purpose of phase three is to commercialize the results of phase two and requires the use of private-sector or non-SBIR federal funding.

Another funding option through the Small Business Administration's technology program is the STTR program. This program focuses on the encouragement of public-private partnerships. Similar to the SBIR program, the STTR program provides funding to small businesses; however, those companies must have a US-based research institute with which it collaborates on the research. The goal of this program is to encourage US universities to make the move from basic to applied research. Eligible companies follow the same criteria as SBIR firms. The nonprofit research institutes must be US based and either a university or college, a federally funded research center, or a nonprofit research organization. Five agencies fund the STTR program: the Departments of Defense, Energy, and Health and Human Services; NASA; and the NSF. Similar to the SBIR, STTR is a three-stage program: In phase one, firms are awarded up to $100,000 for a feasibility test. In phase two, the programs provide up to $750,000 for two years to examine potential commercialization. In phase three, companies must seek external funding.

### Government Versus Industry Funding

One of the major changes in the United States during the 1980s associated with the Bayh-Dole Act was the shift of national R&D funding priorities. Industry became a major provider of R&D funds (Rahm, Kirkland, and Bozeman 2000; Mowery 2004; NSF 2013). Furthermore, universities increased their share of national basic research funding from less than 2 percent in 1953 to 10 percent in 2010.[4] This increase trend is confined to basic research. The university share of total national R&D only grew from 1 percent in 1953 to 3 percent in 2010. However, industry leads in applied

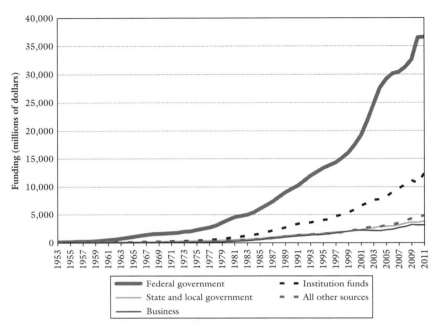

FIGURE 3.1  R&D in universities by funding source
SOURCE: National Science Foundation (2013).
NOTE: Federal R&D spending has grown steadily over the years.

research, which increased its support from 35 percent to 52 percent of applied research (Rahm, Kirkland, and Bozeman 2000; NSF 2013).

As Figure 3.1 shows, when we examine the direct government contribution to university research (not national R&D), we find that federal R&D spending has grown steadily over the years, and as of 2010 it provided about 63 percent of total academic research funding (Mowery 2004; Harrison and Weiss 1998; NSF 2013). Hence, while industry's contribution grew over the years and peaked in 1997 at 7.4 percent, the oft-made claim that industry funding of university research is the outcome of decline in government funding is false. Moreover, in 2010 industry support for academic R&D stood at 5.3 percent (Hatakenaka 2002; Harrison and Weiss 1998; NSF 2013).

*Universities' Response to Changes in S&T Policy*

At the beginning of the twentieth century, universities were quite ambivalent about patenting and commercialization: many refused to deal with their faculty's inventions. Such was the case in Wisconsin in 1924, when the University of Wisconsin refused to be involved in the many inventions of Professor Harry Steenbock. To address the problem, Steenbock created, with the help of several alumni, the Wisconsin Alumni Research Foundation. The foundation filed his patents and licensed them, paying the university back via an annual grant. During the 1920s and 1930s, several other universities followed Wisconsin's example and created their own institutions to manage the commercialization of university intellectual property (Mowery and Sampat 2001a).

The US government's support of university-industry collaboration in the form of the Bayh-Dole Act created an increase in disclosures and patents. Before 1980 there were fewer than 250 university patents per year, and most discoveries were not commercialized. However, the Bayh-Dole Act affected universities' disclosure, patenting, and licensing output. According to the Association of University Technology Managers (AUTM), between 1991 and 2011, annual invention disclosures by US universities increased by 300 percent (to 19,732), new patents filed increased more than 800 percent (to 17,856), and new licenses and options executed increased more than 400 percent (to 5,398) (AUTM 2013).

As Figure 3.2 shows, this growth generated the need for the creation of many professional technology-transfer offices, even at universities not as enchanted with industry (Feldman and Breznitz 2009; Shane 2004).

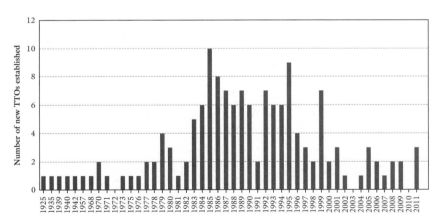

FIGURE 3.2 Growth in the establishment of technology-transfer offices after Bayh-Dole

SOURCE: AUTM (2013).
NOTE: The number of TTOs grew rapidly after Bayh-Dole in 1980.

Before 1980 there were fewer than twenty technology-transfer offices (TTOs); this number grew rapidly after Bayh-Dole, and by 2004 most universities had TTOs (Sampat 2006). These offices vary, as universities do, by size, financial, and professional capabilities. Most technology-transfer offices require faculty to disclose any invention that has patent potential.[5] Each invention goes through an evaluation process, during which the office of technology transfer decides whether it is worthwhile for the university to patent the invention.

One of the ways by which universities affect local economic development is the creation of new firms based on research done in university laboratories. Such firms are known as university spin-outs or spin-off firms. Between 1994 and 2011, the total number of spin-outs by US academic institutions grew from 175 per year to 617 per year, and the number of institutions that spun out companies grew to 77 percent of the total institutions (AUTM 2009, 2013; Shane 2004; Bartik 1985).

In addition, many US universities added supplemental services and channels to increase university-industry linkages and contributions to local economic development. Some universities created collaborative research centers, incubators, and research parks, and at some, companies pay an annual fee to be part of a research consortium that provides information on the latest research in different technological fields (Breznitz, O'Shea, and Allen 2008). Breznitz and Feldman (2012) identified these activities as additions to the traditional roles of universities in teaching and research, categorizing them as knowledge transfer, policy development, and economic initiatives.

Knowledge transfer includes a wider scope of university initiatives that contribute to economic growth. Some of the basic technology commercialization activities, such as licenses and university spin-out companies, have been studied in depth and quantified to demonstrate how universities contribute to local economic growth (Druilhe and Garnsey 2004; Etzkowitz 1995; Felsenstein 1996; Jaffe, Trajtenberg, and Henderson 1993; Jensen, Thursby, and Thursby 2003; Keeble 2001; Miner et al. 2001; Minshall, Druilhe, and Probert 2004; Cantwell 1987). By and large, technology commercialization, formalized by Bayh-Dole, demonstrates a paradigm shift in which universities seek to gain greater relevance.

The availability of interdisciplinary knowledge and experts places universities in a unique situation in which policy development becomes one of the most common roles for universities with regard to promoting economic development. Many academics are individually involved in policy research and development. Therefore, it is not surprising that some universities choose to form task forces that improve the local economies by means of policy decisions.

Last, in the past century universities have taken more economic initiatives. These activities are categorized by Breznitz and Feldman (2012) into four programs: (1) workforce development, which involves workforce education and training; (2) partnership development, in which the university is a tool through which businesses, not-for-profits, and government agencies can connect to debate and collaborate on issues that, in turn, contribute to economic growth; (3) community development, in which universities work with local community groups to invest directly in public education through local schools; and (4) real estate development, to improve their adjacent neighborhoods.

## UNITED KINGDOM

In the United Kingdom, the view of research and higher education as a means to promote local economies started in the early 1900s with the establishment of what were known as the Redbricks universities. Unlike the land-grant universities in the United States, these universities were privately funded institutions. Unlike the older universities, Oxford and Cambridge, these were noncollegiate universities, which focused on providing professional skills for their students. The government role as funder, promoter, and user of university research did not develop until the 1950s. It was not until 1949 and the formation of the National Research Development Corporation (NRDC)—and increasingly so after the 1965 Science and Technology Act—that the government directly funded research in a systematic way. The growth of industry and industrialization processes in the United Kingdom led to a greater need for skilled employees to secure a suitable knowledge base. Universities became more cognizant of the importance of research (Germany's dominance over Britain was viewed as a direct result of German universities' research capacity), but at first, government support for this research was slow to emerge (Rahm, Kirkland, and Bozeman 2000).

### Science and Technology Policy

The UK university system was created in waves. The first wave included Oxford and Cambridge, established in the thirteenth century in England,

and Scotland's four universities, starting with St. Andrews in 1411 and followed by Glasgow, Aberdeen, and Edinburgh. Additional universities were not created in the United Kingdom for another six hundred years, when universities such as London and Durham were established in the 1830s. The Queen's University of Belfast was established in 1845, and the Federal University of Wales in 1893. In the nineteenth century, universities in the United Kingdom were still viewed as institutions that provided liberal education rather than applicable skills; thus, education there was made available only to a small portion of the population—the elite—who did not need to acquire a particular skill to provide for themselves. The third wave of universities occurred in the early twentieth century with the establishment of the civic universities in Leeds, Liverpool, Birmingham, Sheffield, Bristol, Reading, Nottingham, Southampton, Hull, Exeter, and Leicester. These were collectively called civic universities, as they were founded to bring the benefits of higher education to provincial life. The Redbricks (Birmingham, Liverpool, Leeds, Sheffield, Bristol, and Manchester) were involved with industry, so their creation was justified as a contribution to the local economy (Rahm, Kirkland, and Bozeman 2000). The Redbricks were the first attempt in the United Kingdom to promote university-industry relationships. However, the initiatives were all private.

Over the years, a few programs were established by the UK government to help meet the universities' funding needs. For example, the University Grants Committee, established in 1919, funded full-time faculty to allow them to work solely on research. Also, the Medical Council, established in 1920, and the Agriculture Council, established in 1931, supported research in their fields. Most of the social sciences, however, still relied on private foundations.

Although World War II had a profound impact on the United States' S&T policy, it had only a marginal impact in the United Kingdom. This was because academics in the United Kingdom were taken out of university labs and placed in military ones; after the war, academics went back to their academic research, losing any continuity with what they did in the military labs. Also, many research activities took place outside of the country, mostly in the United States (Hatakenaka 2002). It was only in 1949, with the formation of the NRDC, that a connection was made between university inventions and national economic promotion. The NRDC focused on commercializing public research, mostly research that had been done during World War II. However, even the NRDC did not help with the slow response of the government to fund research.

The Robbins Report in 1963 was the first to provide a national frame-work for university contributions to support the demand for local learning and local economy. As chair of the Committee on Higher Education from 1961 to 1964, Professor Lionel Robbins spearheaded a heated debate in the United Kingdom with his remarks on the need for educational availability to all. According to Robbins, "All young persons qualified by ability and attainment to pursue a full-time course in Higher Education should have the opportunity to do so" (Great Britain House of Commons 1963, 49).

Following the report, the government created another nine universi-ties, known as the "new" or "Greenfield" universities, including Sussex and Warwick. Another recommendation by the Robbins committee was that the existing colleges for advanced technology, created between 1956 and 1962, should become universities awarding their own degrees. This created a wave of technical universities, which shaped their teaching to accommodate industry, creating courses that allowed students to have in-ternships as part of earning their degrees. Twenty-three universities were created in the 1960s, and an additional forty in the 1990s, when the poly-technics, created in the 1960s for students unable to attend universities and seeking technical degrees, were given university status as a result of the Further and Higher Education Acts of 1992.

Formal support and use for university research in industry in the United Kingdom continued with the creation of the 1965 Science and Technology Act, by which a formal structure for the national research councils was created. This structure also provided support to the previ-ously established medical and agriculture councils, establishing the Envi-ronmental and Social Science Councils and the Science Research Council. Although their names have changed—the Science Research Council is now the Science and Engineering Research Council, and the Social Science Re-search Council is the Economic and Social Science Research Council—the councils and their structures still exist.

The current dual-government funding structure of British higher edu-cation was created through two sources of funding: the Higher Education Funding Councils (the replacement for the University Grants Commit-tee) and the Research Councils.[6] The funding council grants include those from the Higher Education Funding Council for England, the Higher Ed-ucation Funding Council for Wales, the Scottish Further and Higher Education Funding Council, the Training and Development Agency for Schools, and the Department for Employment and Learning in North-ern Ireland, and research grants and contracts, including research councils covered by the Office of Science and Technology: Arts and Humanities Research Council, Biotechnology and Biological Sciences Research Council, Engineering and Physical Sciences Research Council, Economic and Social Research Council, Medical Research Council, Natural Environment Research Council, and Science and Technology Facilities Council.

Starting in 1980, the funding councils used the Research Assessment Exercise to allocate funding on the basis of the number of students and research quality. This assessment is done every four to five years (Rahm, Kirkland, and Bozeman 2000). While the higher education funding councils provide funding for basic research, at the university's discretion, the research councils' funding is directed, similar to that of the NSF or NIH in the United States, to specific research projects (Adams and Bekhradnia 2004). As Figure 3.3 shows, government funding of research, including funding councils, research councils, and government departments, for UK universities is 69 percent.

In 1981, the NRDC was combined with the National Enterprise Board to form the British Technology Group. Until 1985 the rights on university inventions funded by the research council were held by the British Tech-

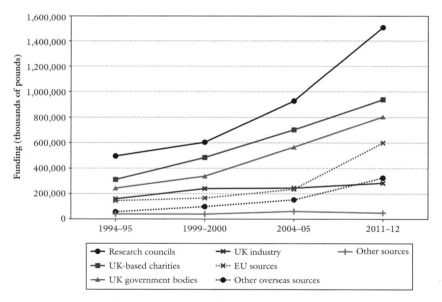

FIGURE 3.3  Research funding at UK universities, 1994–2012
SOURCE: Higher Education Statistics Agency (2012).
NOTE: UK government funding for research has grown to provide 69 percent of total funding.

nology Group. But even after 1985, when universities could file to own those rights, most did not. Many claimed that the fact they had to file reports with the research councils on their commercialization activities discouraged them from claiming the inventions. In 1991 the British Technology Group was privatized, and hence the rights for inventions were returned to the universities. Without an official act like the Bayh-Dole in the United States, universities in the United Kingdom held the rights to their nationally funded research. This process allows universities to engage directly with industry, and as a result, industry funds a certain proportion of university research. However, as is evident from Figure 3.3, industry funding of research went down from 7 percent in the 1990s to 6.3 percent in 2011, which is similar to the percentage of industry sponsored research in US universities (Higher Education Funding Council for England 2003; AUTM 2013).

The UK government supports university-industry relationships through the creation of programs encouraging universities and businesses to collaborate on research. The earliest program, the Teaching Company Scheme, was created in 1975 to provide businesses with academic support through the employment of a graduate student in a company's research project. Currently, the program is part of the Knowledge Transfer Partnership. The program defines itself as follows: "Knowledge Transfer Partnerships is a UK-wide programme helping businesses to improve their competitiveness and productivity through the better use of knowledge, technology and skills that reside within the UK Knowledge Base" (Department of Trade and Industry [DTI] 2005).

The Knowledge Transfer Partnership connects a business with an academic research institution and an associate (usually a postgraduate student). Over the years, success has come through continuities and evolving relationships. The higher education institutes saw an increase in new research projects and papers, and the business partners saw an increase in revenues. The program is based on the academic expertise of the higher education faculty, who work with a business to assist in its future development. A team with members drawn from both the company and academic institutions manages it (DTI 2005).

While the Knowledge Transfer Partnership sponsors close-to-market research, the second program, LINK, created in 1986, sponsors innovative and basic research. In 2004 LINK was absorbed into the Technology Strategy Board. Accordingly, its name was changed to Collaborative Research and Development Grants:

> Collaborative Research & Development (R&D) is a support product, provided by the Technology Strategy Board. It encourages businesses to team up and work with each other and with those in the research community to develop innovative products, processes and services in specific areas of technology or technology application. (Office of Science and Technology 2005)

These grants allow the development of basic research to achieve market viability. Government provides up to 75 percent of R&D costs.

The third program that promotes university-industry relationships through research is Cooperative Awards in Science and Engineering. This program provides joint supervision of students by a member of an eligible academic institution and an employee of the cooperating industrial company. Students spend part of their training period working within the company. There is also the industrial Cooperative Awards in Science and Engineering, in which the industrial partner defines the research topic and takes the initiative in establishing a link with an eligible academic institution (Hatakenaka 2002). Since 2008, many of these programs are managed through the Technology Strategy Board, the national innovation agency in the United Kingdom. The agency has since then launched or is in the process of creating new programs such as the SBIR (similar to the US one) and the technology innovation centers, both of which will have an impact on university-industry partnerships.

After the Robbins Report of 1963, the only major report on the future of the British higher education system commission by the government was the 1997 Dearing Report. That report, "Higher Education in the Learning Society," was the result of the 1990s crisis in the UK higher educational system, where policies that increased the number of students reduced the funding per student (as can be seen in Figure 3.3). The amount universities were allotted to spend on teaching was reduced by 50 percent, and funding for infrastructure and research was also cut.

The report is actually a collection of reports, resulting in ninety-three recommendations for general university funding, students' fees, and teaching in higher education. Funding was a significant part of the report's recommendation. Some of the Dearing Report recommendations for university involvement on regional and community levels can be seen in the university-industry government initiatives as further historical factors that influence the technology-transfer capability of UK universities. One kind of incentive was in the form of financial allocation to promote universities' technology transfer and commercialization.

The results of the Dearing Report (1997) can be seen in the creation of the following programs, which in many universities, such as Cambridge, resulted in the creation of many separated offices. The Higher Education Reach Out to Business and the Community Funding Program of 1998 provided £20 million ($29.7 million) to assist in the establishment of activities such as technology-transfer offices. The University Challenge Fund, initially a collection of fifteen funds that allowed access to thirty-seven institutions, was set up with an investment of £45 million ($67 million) by the government (contributing £25 million, or $37 million), the Wellcome Trust (contributing £18 million, or $26.7 million), and the Gatsby Charitable Foundation (contributing £2 million, or $3 million) to assist universities in turning research projects into viable businesses. The government continued its support of university-industry relationships by creating the Science Enterprise Challenge. Launched in 1999, the Science Enterprise Challenge provided £28.9 million for the creation of twelve centers and another £15 million to increase the number to sixty centers, supporting the commercialization of science and technology. The science enterprise centers are described thus:

> UK Science Enterprise Centres have a government mandate to change the culture within universities. They form a part of the UK government's strategy to add enterprise to Higher Education's mission alongside teaching and research.
> UKSEC forms a network of 13 local centres representing over 60 higher education institutions. Minister for Science, Lord Sainsbury explained the role of the UK SECs as "catalysts of cultural change in UK universities to make them more relevant to business and to enhance the universities' contribution to growth in the economy, employment and productivity." (UK Science Enterprise Centres 2005)

The Higher Education Innovation Fund, created in 2001, builds on the existing program of Higher Education Reach Out to Business and the Community, providing funding "to support universities' potential to act as drivers of growth in the knowledge economy" (Minshall and Wicksteed 2005, 4). Thus, through these programs, the UK government provided funding to support technology-transfer and commercialization programs at UK universities.

In 2002, a time when the United States was an obvious leader in innovation and university-industry collaboration, the United Kingdom finally engaged in a real debate on the role of universities in the development of its economy. In November 2002, Chancellor Gordon Brown and Trade and Industry Secretary Patricia Hewitt asked the former editor of the *Financial Times*, Richard Lambert, to lead an independent review of "how we can boost the UK economy by strengthening the long-term links between business and universities" (DTI 2005). The Lambert Review, published in 2003, provides an assessment of university-industry relationships in the United Kingdom:

> There is much more to be done. Universities will have to get better at identifying their areas of competitive strength in research. Government will have to do more to support business-university collaboration. Business will have to learn how to exploit the innovative ideas that are being developed in the university sector. (Lambert 2003)

According to the report, business-university collaboration needs to be strengthened, and a more realistic view should be taken of the limited ability of universities to make money directly from intellectual property. Government should support university departments doing work that industry values by funneling funds through the regional development agencies, which can play greater roles in university-industry relations.

The Lambert Review found that business demand is the main caveat in university-industry relationships. Even though UK universities are highly ranked in R&D performance, UK businesses do not value university research—especially not when other multinationals would choose to locate their facilities in close proximity to centers of excellence, in and outside their home country. Interestingly, the review did not support legislation like the Bayh-Dole Act in the United States (Lambert 2003).

### *Universities' Response to Changes in S&T Policy*

The combined result of government activities starting in the late 1980s and the beginning of the 1990s, such as the privatization of the British Technology Group in 1991, university-industry relationship programs such as LINK and Cooperative Awards in Science and Engineering, and governmental funding such as Higher Education Reach Out to Business and the Community, can be seen in the growth of university patents in the United Kingdom (see Figure 3.4).

Unlike the Bayh-Dole Act in the United States, the 1991 UK act was not clear with regard to university taxation. Thus, many universities in the United Kingdom have created wholly owned private companies to manage licensing and patenting revenues. In 1997, more than half of the higher education institutions in the United Kingdom reported ownership of such institutions to exploit their intellectual property (Hatakenaka 2002).

Another result of the government programs is the evaluation, reorganization, and creation of TTOs in the United Kingdom. The early technology-transfer offices in the 1960s and 1970s established policies regulating the

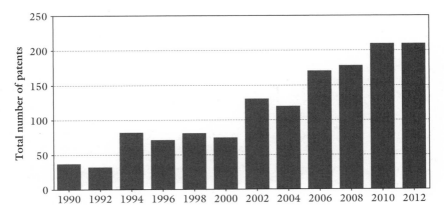

FIGURE 3.4  UK universities' patents, 1990–2004
SOURCE: European Patent Office (2006, 2013).
NOTE: UK universities' patents have grown as a result of government funding.

ownership of their technology as well as the creation of industrial liaison functions. In some TTOs, students and postdoctorate fellows were required to license their inventions even though they were employees of the university. Many of these first offices were small—one survey found an average of six full-time employees—although in some universities the offices had a greater number of employees with better technical expertise and experience.

With the changes of the 1980s and 1990s, however, technology-transfer offices became prominent. Thus, more importance has been given to employee background and to the number of employees, which makes the research support function just as important as finance or registry. Still, there is much variation in technology-transfer offices. In some, legal or patent specialization of the office is sourced out, and in others, it is in-house. Some universities have privatized their technology-transfer offices, thus allowing organizations to finance their own activities and not be restricted by university finances (Rahm, Kirkland, and Bozeman 2000).

## NATIONAL-LEVEL SUMMARY

Our review of the national contexts reveals that until the 2000s, universities in the United States and the United Kingdom operated in different arenas and were influenced by different historical and legislative factors.

In the United States, university-industry relationships were viewed favorably and were financially supported by the federal government beginning in the 1950s, but in the United Kingdom, support did not start until the 1990s. Thus, industry in the United States has used and relied on academic support for a longer time than industry in the United Kingdom. There, industry is only beginning to realize the advantage of university-industry collaborations and the potential of noncommercial basic research. Yale University has been fortunate to operate in an environment that promotes collaboration with industry and commercialization of university research. Cambridge, however, was a medieval university that has had to adjust its role to become an engine of economic growth.

As is seen in Figures 3.5 and 3.6, the two countries' funding for higher education research shows heavy government reliance. The US federal government provided 60 percent of the total funding for academic R&D, in the amount of $36 billion in 2011. While the federal government is still the largest contributor to academic R&D, its share of the total R&D expenditure is on a continuous decline, from almost 70 percent in 1972 (Harrison and Weiss 1998; AUTM 2013; NSF 2013).

The UK government provides 52 percent of the total higher education funding for research in higher education, in the amount of £2.314 billion. By 2009, the United States was spending 2.6 percent of its gross domestic product (GDP) on higher education, whereas the United Kingdom was

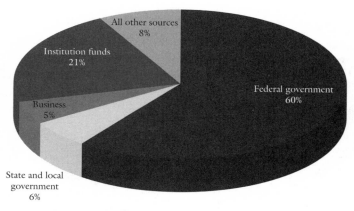

FIGURE 3.5  Sources of academic R&D in the United States, 2011
SOURCE: National Science Foundation (2013).
NOTE: The US federal government provided 60 percent of the total funding for academic R&D.

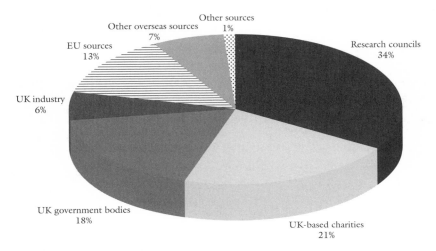

FIGURE 3.6  UK research grants and contracts income of HEIs, 2011–12
SOURCE:  Higher Education Statistics Agency (2012).
NOTE: The UK government provides 52 percent of the total funding for research in higher education.

spending 1.3 percent of its GDP on higher education. Furthermore, in examining the two countries' total research spending (not just in higher education), we find that in 1981 the United Kingdom and the United States spent about 2.4 percent of their respective GDPs. While the United States continued to invest in research and reached 2.9 percent of the GDP, in 2009 the United Kingdom share declined to 1.86, lower than its investment in 1981 (Moulaert and Sekia 2003; Breznitz and Ram 2013).

## CHAPTER FOUR

# *Yale University*

Yale University is one of the world's most highly regarded research universities. It has a strong life sciences faculty and extensive research capabilities that fuel the creation of a biotechnology cluster. Although as late as 1993 this cluster had only 6 biotechnology and pharmaceutical companies (compared with Boston, Massachusetts, which had 129 such companies by 1993), by 2004 it consisted of 49 such companies, which grew to 70 companies by 2013. Between 1993 and 2004, Yale University had made important changes that led to the growth of a biotechnology cluster in particular and a positive impact on the local economy in general.

Those changes included technology-transfer changes directed toward local economic development. To improve technology commercialization, Yale conducted extensive changes in policy, organization, and culture. Moreover, the university was only one actor in a region geared toward the biotechnology industry. Changes at the university were complementary to, as well as supported by, the State of Connecticut, the City of New Haven, and local industry.

## EXTERNAL UNIVERSITY FACTORS

Yale resides in the city of New Haven, Connecticut. Total population for the state as of the 2010 census was 3.5 million, and the total population of the city of New Haven was 129,700. Geographically, the city is well situated within the Northeast, about an hour and a half, or 80 miles, from New York City and about two and a half hours, or 137 miles, from Boston,

Massachusetts. There are no direct flights to New Haven, with the local airport providing flights only to Philadelphia.

The New Haven metropolitan area is home to seven institutions of higher learning: Yale University, Southern Connecticut State University, Albertus Magnus College, Gateway Community College, Quinnipiac University, Paier College of Art, and the University of New Haven. Furthermore, and important for our discussion of the development of the biotechnology industry, there are five multinational pharmaceutical companies with research facilities in the region: Pfizer, Bristol-Myers Squibb, Purdue, Bayer, and Boehringer Ingelheim.

Despite its location, size, and history, the city has experienced problems similar to those of much larger US cities. The city's crime rate in the 1990s was higher than the US national average, specifically in what is known as the Hill neighborhood, Science Park, and the area near the Ella Grasso Boulevard. In 1994, New Haven was described as "a war zone of poverty, crime and drugs, as frightening as any city in America" (Sedgwick 1994, 182). The sense of crisis in the city began intensifying in the 1980s. In 1989, thirty-four people were murdered in New Haven, the city's highest homicide rate in recent history. Fueled by the crack cocaine trade, rival gangs fought violent turf battles all over the city (Ball 1999). Safety and security were particular issues for local high-tech firms when considering whether to locate in the city center. Seemingly mundane issues, such as street lighting, provision of sidewalks, and parking, became important factors in deciding whether to remain in the city.

Despite its favorable location and the existence of other research institutes and multinational companies, the city did not benefit from rounds of high-tech growth before 1996. Until then, the State of Connecticut did not have any incentives for the development of the biotechnology industry. In fact, two biotechnology companies, Exilexus and Gene Logic, chose to leave the region, blaming lack of state support in the form of laboratory space and funding for young companies. Both of the firms are considered an industry success. Gene Logic moved to Maryland and was bought by Ocimum Biosolutions, an Indian life science company, for $10 million in 2007. Exilexus moved to San Francisco and completed its public offering in 2000; it is now traded on Nasdaq.

In 1998 things began to change. Based on the work of 125 Connecticut business leaders, six industries were identified as key sectors for Connecticut's economic development, including biotechnology. The state launched an industry cluster initiative under the Department of Economic and

Community Development, basing its efforts on Michael Porter's (1990, 71) cluster methodology, which incorporates four factors to achieve national competitiveness:

> (A) Factor conditions—the nation's position in factors of production, such as skilled labor or infrastructure, necessary to compete in a given industry. (B) Demand conditions—the nature of home demand for the industry's product or service. (C) Related and supporting industries—the presence or absence in the nation of supplier and related industries that are internationally competitive. (D) Firm strategy, structure, and rivalry—the conditions in the nation governing how companies are created, organized, and managed, and the nature of the domestic rivalry.

The first to launch in 1998 was the bioscience cluster, as one of many collaborative efforts to promote local economic development in the region. To further its investment in the local economy, the State of Connecticut decided to provide assistance to the industry in several ways: (1) funding through the creation of Connecticut Innovations, (2) active civil service agency through the office of bioscience, and (3) tax incentives.

The first organization created to support biotechnology in the region was Connecticut Innovations (CI). Created by the legislature in 1989, CI was charged with investing in local companies to enhance economic development. It was originally funded by the state, but since 1995, the company has financed its equity investments solely through its own investment returns rather than taxpayer dollars. In fact, CI became the state's leading investor in high technology. The mission of the organization is to make "equity investments in emerging Connecticut technology companies; providing essential, non-financial support to entrepreneurs; and conducting initiatives that address specific needs of Connecticut's technology sector" (CI 2003).

Connecticut Innovations has several ways of investing. Although it is generally an active investor, participating in the creation of a company, writing the business plan, and helping to select the management team, CI sometimes joins in the bridge round or Series A of the financing process.[1] Carolyn R. Kahn, a bioscientist by training, was appointed to lead CI's investments in bioscience in 1998.[2] Two major sources of funds available to the local biotechnology industry include the Connecticut BioSeed Fund, a $5 million fund administered by CI, which provides seed capital to address the initial financial needs of young Connecticut companies. Another, larger, fund is the Bioscience Facilities Fund, a $60 million fund used to underwrite the development of incubator and lab space. As simple as this may seem, Connecticut in general and the area around New Haven and Yale in particular did not have enough commercial space and biotechnology-specific space (e.g., buildings that can hold a wet lab). Moreover, small firms have difficulties in funding their expensive equipment and space. These funds are not available through regular venture capital investments. Hence, the facilities fund played an important role in helping firms move from university labs to the commercial sphere.

The second state initiative, created to promote the bioscience cluster, is the Office of Bioscience. With the second cluster bill in 2001, the State of Connecticut allocated $100,000 to establish the Office of Bioscience within the Department of Economic and Community Development. This office was established to support start-up and existing companies in the region, to provide all the necessary information on conducting business in Connecticut, to bring new and existing out-of-state companies to the region, and to represent the life science cluster of Connecticut in national and international events (interview with one of the office's executives). The Office of Bioscience is one of the environmental factors contributing to the creation of a life science industry in Connecticut.

There are also other factors contributing to the success of technology transfer and university-industry relationships in the region, in the form of Connecticut's tax incentives for the biotechnology industry. These include the 1996 Biotechnology Tax Incentive Package, which provides exemptions from sales, use, and property taxes, as well as a fifteen-year carry-forward R&D tax credit; the 1999 Tax Credit Exchange in which eligible companies that cannot use their R&D tax credits can exchange them with the state for 65 percent of their value; and the Sales Tax Relief, which provides 50 percent and 100 percent exemptions on certain biotechnology industry materials, such as tools, fuels, equipment, and safety apparel.

Besides state initiatives, the industry itself created institutions to promote the industry. The main example is Connecticut United for Research Excellence (CURE), originally an educational organization that in 1998 somewhat unexpectedly became the industry representative. In addition to its educational role, before 1998 CURE had also been a venue for many of the biotechnology companies to meet and discuss issues related to its (highly regulated) industry; this was necessary partly because the industry could not find a voice at the Connecticut Technology Council. In enacting the cluster bill, the State of Connecticut formalized CURE's unofficial role for the industry, making it the overseeing body for the bio-science cluster in Connecticut. The state has also contributed some funds to allow CURE to broaden its educational activities. Later, most of the state funding was used to cover CURE's participation in the Office of Bioscience.

The main activities of CURE are (1) lobbying for the interests of the bio-pharmaceutical industry, specifically seeking to preserve tax incentives for the industry at a time when the state is running a large budget deficit, and also working to develop a qualified labor force for the industry by creating certificate programs in local colleges; (2) educating the public in general and children in particular on the science of biotechnology through the Biobus program, hoping to stimulate interest in studying and working in life sciences; and (3) acting as a conduit for information needed by the industry (e.g., how to manage a laboratory, how to build an animal lab). Under CURE's auspices, local firms' top management, such as their chief executive officers, chief financial officers, and public affairs and human re-source executives, meet quarterly to share information on similar problems and solutions.

Another important player in the creation of the local biotechnology cluster in New Haven is the City of New Haven. The city supported the biotechnology industry by providing it with basic infrastructure and ad-dressing safety concerns, such as adding streetlights and building side-walks, especially in the area adjacent to Science Park on 300 George Street, located in a neighborhood with one of the highest crime rates in the city. Industry officials in general and in the biotechnology industry in particu-lar were positively impressed by the city's contribution:

> The city administration is doing more than I expected. They support business. This area here was not a very good part of town; they renovated the area, put streetlights. We did not have a single incident of assault. The city government should take credit for that. The police force is very responsive. (Interview with biotechnology executive)

Reviewing the history and environment in the region in which Yale University operates, we find that the general support for university-industry relationships and the biotechnology industry in particular began only in the late 1990s. However, the regional support was a comprehensive and coordinated process among many of the institutional players in the region, where Yale University was only one of the actors. Importantly, the university had a pivotal role of leadership and organization.

INTERNAL UNIVERSITY FACTORS

By the early 1990s Yale University found itself in an academic and physical crisis. The university was located in a region that following the high-tech boom of the 1990s had no high-technology cluster. Moreover, crime levels were high to the point that the university feared it would lose students and faculty. These reasons, together with an operating deficit, created a sense of crisis at Yale. As expressed by Richard C. Levin, Yale's president from 1993 to 2013:

> By the time I came in as president, it [economic development] was a necessity in the sense that our position in attracting faculty and students had begun to drop, with people citing the environment as a reason for declining offers of jobs or admission. I enunciated the importance of local development, but it wasn't even controversial. By my acceptance speech people were ready for it. Hence, recruitment of faculty and student was part of the reason. And Safety. There was a lot of urban crime at that stage. (Interview with Richard C. Levin, August 12, 2013)

Thus, the university chose to approach the crisis via cultural, organizational, and policy changes directed specifically at local economic development, technology commercialization, and firm formation.

### *Organization*

The way a university technology-transfer office is organized has been found to be an important factor in the ability of the university to patent and license its technology (Link and Scott 2005; O'Shea et al. 2005). Yale is a private university that fully funds its technology-transfer office and its activities. One hundred percent of the office's expenditures are funded directly from the administration and are not royalty dependent. The office collaborates but is administratively separated from the office of sponsored research. Moreover, the Office of Cooperative Research employees are educated and experienced in technology-transfer commercialization. Thus, the office employees have knowledge, background, and experience similar to that of the staff in many of the firms with which they collaborate. Industry's view of and relationships with a university technology-transfer office reflect the professionalism of that office. Yale University made a choice to invest in its technology-transfer process and thus chose to recruit the most qualified people to work there. As a result, the biotechnology industry in New Haven views Yale University as an important partner with which it can work. Affirming organizational change theories that view successful changes based on new personnel with leadership and marketing skills all can be found in the changes implemented at Yale. What follows is a look back at the history of these changes.

In a period in which many universities contributed to the development of, and operated within, high-technology and biotechnology clusters, Yale had none. By 1993, it was clear to many at the university that this situation was far from ideal. The lack of a local industry, particularly in life sciences, which is the university's strength, raised concerns among faculty, students, and administrators:

> What was happening was the university was starting to become concerned that it would detract from our ability to compete, to attract the best and brightest students, the best and brightest faculty, et cetera, if we didn't do something about it. . . . First and foremost it was all about enhancing our reputation as a university, and two things come from that. One is our ability to attract and retain the best and the brightest faculty and students, and the second is to diversify the regional economy. Those were probably the principal reasons, and we weren't against making money, but we weren't making a lot at the time. It really wasn't the principal motivator; it really was about our reputation. (Interview with Yale director)

Yale was concerned that the lack of industry presence and collaboration would harm its ability to attract and retain star scientists and bright students, thus damaging university research and reputation.

In addition, the city of New Haven was not considered a safe place for Yale's students. This became acute with the shooting and death of a Yale undergraduate in 1990. As a university within a city, Yale had to fight against local crime to ensure the safety of its students by working with the city of New Haven to revitalize the downtown area and assist its employees in purchasing homes in the city. To make New Haven safe and more attractive to potential students and faculty, a $2 million project in 1993–94 put streetlights on nearly every corner of the Yale campus and installed an emergency campus phone system and electronic entryways.

At the same time, Yale University had an operating deficit that, by 1993, had reached $14.8 million. This deficit arose in 1991 under the presidency of Benno C. Schmidt Jr. In response, the university addressed this problem through budget cuts and eliminating several departments, and hence had no deficit by 1998.

Thus, Yale University in the early 1990s was facing several crises, which it chose to address through organizational changes. Changes included the technology-transfer office and the Office of Cooperative Research (OCR). Studies show that recognizing failure and attempting to create positive outcome can justify new forms of organization and resource allocation (Tolbert and Zucker 1996; Greenwood and Hinings 2006).

As we know from existing studies, the arrival of new people to an organization as well as their connection to an external network affects an organization (DiMaggio and Powell 1983; Greenwood and Hinings 2006; Schein 1985; Whelan-Berry, Gordon, and Haining 2003; Dacin, Goodstein, and Scott 2002; Bandura 1977). Hence the appointment in 1993 of a new president at Yale, Richard C. Levin, an individual with an economic background who had concerns for the future of Yale, was the first step of the organizational change. Similar to many new leaders, Levin drove a comprehensive organizational change at the university that included departments such as the OCR and the Office of New Haven and State Affairs.

The sense of crisis at Yale allowed Levin to implement a vast social, cultural, and economic development change. His changes supported theories that moments of crisis allow room for cultural change and that economic growth is anchored in the cultural change of that institution. In many cases, firms (e.g., Xerox) choose to change their practices, social relations, and ways of thinking to better compete in the economic market. Thus, "different cultural trajectories create different interpretive and strategic possibilities in the face of the same technical and environmental conditions" (Schoenberger 1997, 221).

The interviewees who participated in this study overwhelmingly agreed that Levin was the catalyst for the change in Yale's attitude toward research, with potential practical applications. Levin (2003) wanted Yale to be a "contributing institutional citizen" with a long-term commitment to the community. By referring to Yale as a contributing citizen, Levin was referring to a broad range of activities at the university, not solely to its role as an enhancer of economic development.

To pursue this vision, Levin led the university in an in-depth study of the activities already performed by Yale in the community, deciding to invest in four areas: economic development, strengthening neighborhoods, revitalizing the downtown area, and improving the city image. As noted by Levin in 2003:

> The first area of focus was economic development. We had considerable faculty strength in the biomedical sciences, but no track record of encouragement or support for the transfer of technology to local businesses.

The second priority was strengthening neighborhoods. Here we believed that increasing the rate of home ownership could improve the stability of neighborhoods and the commitment of residents, and that the university, with 10,000 employees, had the leverage to help. We also believed that as an educational institution, we had human resources that could assist the work of the public schools.

The third area of focus was to increase the safety, appearance, and vitality of our downtown. We believed that this would greatly improve perceptions of the city and also directly benefit the university community, since we are located in the heart of downtown New Haven.

Finally, we focused on the image of the city, recognizing that improvement in its physical and material conditions was not in itself enough to change perceptions of the outside world. We needed to communicate as well. (Levin 2003)

Levin's speech emphasizes the importance Yale saw in its contribution to economic development. To support the focus on economic development, Yale decided to change its technology-transfer organizational structure, hence rebuilding the OCR, which until then had not made a real attempt to create or promote technology transfer from the academic to the industrial arenas.

In 1995, Levin, together with Yale's provost at the time, Allison Richard, who later became vice chancellor of the University of Cambridge, persuaded Gregory Gardiner, a former Pfizer executive, to head the organizational and policy restructuring of the OCR. A former member of the Yale chemistry faculty, Gardiner remembered the earlier lack of enthusiasm at Yale for research with practical applications and was eager to help bring about change. Indeed, Gardiner had left the university in the 1970s because it did not encourage work on research with potential application. Gardiner's expanded mission changed the function of the OCR.[3]

The drivers for the changes Yale implemented are eminent in the OCR's 2013 mission statement:

Foster cooperative efforts to translate academic research into products and services for the benefit of society, support of the broader research and education missions of Yale, and where possible—

- Catalyze local economic development.
- Enhance the reputation of the university.
- Generate revenue for reinvestment in those missions. (Office of Cooperative Research 2013)

In August 1999, Gardiner retired, and Jonathan Soderstrom succeeded him as director of the OCR. Similarly to Gardiner and following the concept of a professional technology licensing office, Soderstrom brought his vast experience in the field, having served as director of program development for Oak Ridge National Laboratory as well as director of technology licensing for Martin Marietta Energy Systems.

The OCR's activities were characterized by active promotion of commercialization of research on a local level, not merely passive acceptance. For example, since lack of funding was an issue for many of the university spin-outs, during 1996 and 1997 the OCR established direct contacts with local venture capital firms. As indicated in an interview with a Yale administrator, the university contacted the local venture capital industry to inform them of the investment potential in Yale's firms:

> We have all kinds of venture capital. One of the dirty little secrets is that although Boston thinks of itself as a major financial capital, we've got one that's even bigger. It's called Stamford-Greenwich. When there was no state income tax, all the bankers used to live in Stamford-Greenwich, not in New York City. So they all are still there, and that's where they have their finance companies.[4] (Interview with Yale administrator)

Yale contacted the venture capital firms with the goal of not only persuading venture capital firms of the relevance of university technology but also of convincing them of the importance of creating new ventures in New Haven. Their efforts in seeking appropriate investors eventually paid off, and in 1998, after two years of work, the first round of financing was concluded with $20 million for five companies. Table 4.1 provides the total venture capital invested in Yale firms by 2012.

An equally important problem was the lack of appropriate infrastructure, such as laboratory space for new business ventures and urban amenities to make New Haven attractive to mobile scientists and academics. To assist in the development, President Levin used Yale's ability to recruit top talent, and in 1998 he convinced Bruce Alexander, a prominent figure in urban regeneration, to join Yale's Office of New Haven and State Affairs.[5] As explained by a Yale official:

> It became clear that there's no better person to kick out the economic development kind of mission that Yale would like to have than a guy like Bruce, so Rick [President Levin] convinced Bruce to take it on full time. It's one of those things where you sit around going, "It's nice that everyone wants to do this," but how many people are going to be able to tap a guy like Bruce Alexander to be their economic development guru, the guy who redeveloped the Harborplace in Baltimore, the guy who did South Street Seaport in Manhattan? It makes us all look smart, but it's what a university like Yale can do.

TABLE 4.1 Yale biotechnology companies by venture capital funding

| Name | Year founded | Location | Total VC funding |
|---|---|---|---|
| Enzo Biochem | 1976 | NY | $14,000 |
| Alexion Pharmaceuticals | 1991 | CT | $5,800,000 |
| ARIAD Pharmaceuticals | 1991 | MA | $34,300,000 |
| Genelabs Technologies | 1993 | CA | $1,500,000 |
| Neurogen | 1993 | CT | $85,750,000 |
| CuraGen | 1993 | CT | $1,179,000 |
| Exelixis | 1995 | CA | $34,000,000 |
| TransMolecular | 1996 | AL | $43,840,000 |
| Genaissance Pharmaceuticals | 1997 | CT | $26,560,000 |
| Phytoceutica | 1998 | CT | $14,000,000 |
| Molecular Electronics | 1999 | SC | $7,500,000 |
| Agilix | 1999 | CT | $18,810,000 |
| Molecular Staging | 1999 | CT | $35,760,000 |
| Ambit Biosciences | 2000 | CA | $172,990,000 |
| Achillion Pharmaceuticals | 2000 | CT | $193,840,000 |
| Protometrix | 2001 | CT | $4,520,000 |
| Archimex | 2001 | MA | $3,443,700 |
| Kemia | 2002 | CA | $68,890,000 |
| VaxInnate | 2002 | CT | $83,760,000 |
| Rib-X Pharmaceuticals | 2002 | CT | $87,710,000 |
| Aureon Biosciences | 2003 | NY | $40,170,000 |
| Proteolix | 2003 | CA | $117,390,000 |
| Iconic Therapeutics | 2004 | GA | $4,630,000 |
| HistoRx | 2004 | CT | $13,120,000 |
| Access Scientific | 2004 | NY | $15,690,000 |
| Applied Spine Technologies | 2004 | CT | $47,000,000 |
| Marinus Pharmaceuticals | 2004 | CT | $49,810,000 |
| Axerion Therapeutics | 2009 | CT | $2,400,000 |

SOURCE: VentureXpert database, Thomson SDC Platinum Version 4.0.3.1.
NOTE: This list includes only the companies that had venture capital information reported through VentureXpert (Thomson SDC Platinum Version 4.0.3.1).

With the recruitment of Alexander and his vast knowledge of economic development, the OCR, with the Office of New Haven and State Affairs, set out to build laboratory space close to Yale's scientists. Accordingly, the university attracted two developers, Winstanley Associates and Lyme Properties. These developers, though they were not financially involved with the university, had experience in building labs. Winstanley bought the vacant headquarters of the telephone company on George Street, and Lyme took over the development and management of the failing Science Park on the north campus. The university and the city had been trying for years to develop a science park in that area without success.

At the same time, Yale invested in its downtown properties to make rundown New Haven a safer and more enjoyable city. For example, in its Broadway Street properties, Yale created a mix of both local businesses and national chains and transformed the area into a vibrant shopping area and late-night gathering spot. Both steps, the involvement in the development of science parks and the downtown reconstruction, provide another example of Yale's commitment to local economic development and intertwined relationships among the university, the city, the state, and the local economy.

By 2004, the OCR employed eighteen people, each of whom had five to seven years' experience in industry. As a private university, Yale can offer competitive salaries to senior employees recruited from industry. The employee background at the OCR had, and continues to have, a crucial role in the cooperative relationships between local faculty and industry. For example, when asked whether the OCR is skillful and professional, a senior faculty member answered:

> Yes. I do believe that. I don't think there's any question. And I think that communication problem has gotten much, much better. From my biased point of view, I think it's enormously improved, and that they do a very good job at Yale. (Interview with Yale faculty member)

Also, recognizing that 80 percent of patents from Yale were in the biomedical field, the OCR (1998) opened another office at the School of Medicine with four staff members. In addition, in 2007, Yale University opened the Yale Entrepreneurial Institute (YEI), to assist in the development of students ventures. The YEI employs six full-time employees to help with the development of firms and entrepreneurial culture of Yale students, and to connect alumni entrepreneurs with current students.

Not only does the OCR have the staff and expertise; it is also involved with firm creation to an unprecedented level. The office is involved in developing product scenarios, financial projections, and business strategies with the scientists. In many cases, the office is actively involved in building the company, looking for the right management and investors who will succeed in taking Yale's technology to the market. Yale's importance in the creation of companies became clear to one of Yale's professors when he considered moving to a different university:

> I went out there and talked to the head of their tech transfer office. I told him about my company and how Yale contributed to the creation of the company. He said, "I couldn't possibly do anything like that here. We are not permitted, we don't have any resources. What Greg [Gardiner] is doing is unheard of." That's another instance where I began to realize what seemed to be happening at Yale—when it's happening all around you don't realize it, it's just happening very quickly. First, there are no companies, all of a sudden there are a dozen, and then two dozen over a relatively short period of time. (Interview with Yale faculty member)

The level of university-industry involvement within the OCR compared with that of other universities is considered extreme. Even MIT, considered the top university in university-industry relationships, spinning out twenty to thirty companies a year, is not as involved in the creation of companies as Yale. As mentioned many times by Lita Nelsen, director of the MIT licensing office, MIT is confined to patenting and licensing; it does not incubate firms, it does not get involved with company management, and it does not take seats on the firms' boards. The fact that MIT has been operating in the technology-licensing arena for many years and is located in an entrepreneurship region, which seeks and appreciate new spin-outs from the university, reduces the need for MIT to create companies. Hence, the main role of the licensing office at MIT is focused on channeling new innovations from the university (Breznitz 2011).

### Culture

The original Office of Cooperative Research was established in 1982 to deal primarily with licensing and tracking patents. The OCR was established to actively promote technology transfer and the formation of new firms, spun out from the university. However, there were many obstacles facing Gardiner and his team. One of the biggest challenges was to communicate the new priorities and incentive structure to the Yale faculty:

I was asked many times by junior faculty, "if I get involved with new ventures through the OCR, will I still get tenure?" I told the committee [Educational Policy Committee of the Yale Corporation (the Yale Trustees)] that we have to get Yale faculty to understand it is OK. At MIT, history says that this is OK, but at Yale we need a change of culture. (Interview with Greg Gardiner, former director of the OCR)

This remark highlights the difference between policy creation and policy diffusion. While Yale changed its technology-transfer policy and organization, the effects would not take place until the change became widespread.

To achieve this goal of institutional cultural change, the OCR had discussions with departmental chairs and faculty to explain the institutional change and Yale's commitment to individual involvement in economic development. This was an important issue, since at the time Yale had an ongoing lawsuit against a former faculty member who had patented his invention without naming Yale University as an assignee. In 1996 John Fenn, a chemistry professor at Yale from 1967 to 1987, sued the university after it started negotiations over his invention. Fenn claimed that he was emeritus at the time of his research, and hence his research was not subject to the 1980 Bayh-Dole Act, which gave universities and small businesses control over intellectual property developed through government-sponsored research. Yale University recounted, claiming fraud and demanding reassignment of the patent to Yale rather than Fenn. In 2005, the district judge ruled against Fenn, and he was ordered to pay fees and damages (Poppick 2005). This case and the history of Yale, which had also missed patent opportunities such as the transgenic mouse, discussed in Chapter 1,[6] created an environment in which scientists worried about the implication that working on applied research would have on their academic careers and relationships with the university.

A faculty member recalled:

The OCR people came to professors who had records in licensing or industry interaction and asked for ideas to patent and establish companies. They came to my lab; they knew I worked in [X] and [X]. One of the compounds went to [company name]. They also recruited the management for the company. With the change, Yale has become more entrepreneurial, but we are still responsible to our research and students. (Interview with Yale faculty member)

Thus, to convince faculty of the importance Yale was giving to technology transfer and commercialization, OCR representatives approached faculty, such as Frank Ruddle and Tommy Cheng, who worked on applied research and had made important discoveries in the past.[7]

In addition, the arrival of new faculty from institutions like MIT or Stanford, where working on applied research was acceptable and encouraged, as well as young faculty who wanted to have industry relationships, influenced some hesitant faculty to venture into commercialization and even entrepreneurship (Bercovitz and Feldman 2008).

Examining the change in faculty and students at the medical school versus the rest of the university strengthens this point. Between 1995 and 2011, there was a 55 percent increase in the number of total faculty at the medical school, compared with a 35 percent increase at Yale as a whole. Since 1994, student enrollment has grown by 10.3 percent in the medical school, compared with an increase of only 8 percent in the general enrollment at Yale (Office of Institutional Research 2001; Yale University 2011–12). According to Yale's officials, this increase in faculty at the medical school was not planned. Rather, it resulted from growth in research grants and contract funding over the years. Furthermore, in 1995, with the appointment of Michael Merson, MD, as dean of the Yale School of Public Health, there was an increase in student enrollment in the Epidemiology and Public Health Department at the medical school.[8]

The results of the culture change at Yale are evident. Faculty members interviewed at Yale Medical School explained the numerous benefits in having local biotechnology industry:

> Now, for example, we have a company that is occupying some space in the lab. . . . It is good to have them here, because you have interactions with them and transfer of expertise. (Interview with Yale faculty member)

University-industry relationships allow scientific interactions, sponsorship of students, and access to expensive equipment not available at the university, and they expose students to industrial practices.

According to studies, a technology-transfer office that has seen an invention go through the commercialization process and receives royalties has a stronger motivation to continue and work with the existing commercialization process (O'Shea et al. 2005). At Yale, the technology-transfer office has several successful cases of technology licensed by the university. The most famous, and the one that brought the most revenues, was the drug Zerit. In 1988 Yale licensed the compound that became the highly successful drug Zerit to Bristol-Myers Squibb. At the start of the organizational change, this license produced little or no income for Yale, but by 1998 it was generating royalty income of $30 million to $40 million annually. These funds provided affirmation to the importance and value of technology commercialization, and as such they had a positive impact on the commercialization culture at the university. Zerit and others, such as Emtriva and the Lyme Rx Vaccine, provided a good example and motivation for the technology-transfer office to continue its efforts to commercialize.

Existing studies have found that the university's culture influences its ability to collaborate with industry as well as its technology-transfer output—for example, patents, licenses, and spin-out companies (O'Shea et al. 2005; Saxenian 1994; Shane 2004). Yale University was historically known for its culture of noninvolvement in the community in general and with industry in particular. Yale's culture led it to fail in reaping benefits from several important discoveries, such as the transgenic mouse. For many years, Yale was not active in technology transfer, and by 1993 it had spun out only three biotechnology companies. This attitude of noninvolvement in industry changed during the period 1993–96. Starting with the arrival of Richard Levin as its new president, Yale University's culture changed to support local economic development. This change and the arrival of young faculty members from universities traditionally collaborating with industry contributed to a cultural change at Yale. Moreover, successful commercialization has also contributed to the creation of a new culture at Yale that supports technology transfer.

As a result of its efforts to change its attitude toward technology transfer and commercialization, Yale triggered cultural changes. Although Yale did not set cultural change as a direct goal, it became unavoidable. While the university invested in its technology-transfer organization, in rebuilding the downtown area and assisting in the development of laboratory space and making connections to industry, it demonstrated to faculty that the university was determined to support applied research and commercialization. Moreover, while in the past the OCR had focused solely on patenting and licensing, today the OCR's mission is to help the community by transferring academic inventions to the public to enhance the reputation of Yale University and its faculty, and to transfer its academic inventions to the public domain.

Importantly, the fact that the technology commercialization changes at Yale were not limited to policy and organization is crucial to our understanding of the university's ability to make such an economic impact. Without a general change by faculty, who are the main vehicle for innovation at the university, the region would not have seen a growth in university-related start-ups. As mentioned earlier, Yale's startups represent 53 percent of biotechnology firms in the region.

*Policy*

The technology commercialization policy at Yale supports existing studies regarding best practices to promote commercialization in general and spin-out firms in particular. At Yale, royalties are on a sliding scale in favor of the university. Hence, faculty who would like to receive more royalties will need to spin out a company rather than license their technology. This policy was created because of Yale's drive to affect the local economy through the creation of spin-outs and the variety of services offered at the OCR, which are geared toward company formation.

Policy to create and support spin-out companies is another important factor in a university technology-transfer policy. Yale University used its reputation and connected university technology and inventors directly with venture capital companies to create spin-out companies. Moreover, the OCR has a hands-on policy regarding spin-out companies based on university invention and wanted to retain those companies in the region. Yale uses its brand and connections to bring in venture capital firms, to assist companies in building their business plans, and to connect academics and potential management. This hands-on approach is evident in the total number of spin-outs. All but one were created in a period of fourteen years, and, since Yale decided to invest in technology transfer and commercialization, an average of three companies a year are now created.

Another important factor in technology-transfer policy is faculty guidelines with regard to university-industry interactions and commercialization of academic work. The university created new faculty guidelines that focus on differentiating the university lab from any commercial activity. Since faculty cannot hold dual active positions at the university and a private company, Yale's policy allows faculty to focus their attention on either the company or university research, thus resulting in professionally managed spin-out companies.

Following the restructuring of the OCR, the office initiated an examination of the process by which a faculty member disclosed his or her invention to the university. This examination found that there was a need to change the process so as to prioritize the inventions that were most likely to succeed. The examination resulted in a major policy shift by which OCR would seek out new inventions early, examine them quickly, and invest time and effort only in the strongest candidates. In addition, the upgrading of the OCR's practices led to the identification and recovery of more than $220,000 of unpaid royalties from several licenses.

Yale does not have pipeline agreements on research outcomes. Companies can have an option or first right to license the technology from a sponsored research project, but nothing is prenegotiated. Faculty can sit on companies' scientific advisory boards, but they cannot take a full-time position. Faculty can take a full-time position only while they are on leave of absence from Yale:

> They [faculty] can be assigned to advisory boards, they can be consultants, they can do all those things, but subject to our rules on conflict of interest, etc., the only way they can serve in a management or operative position is if they're not full time, so they'd have to be on a leave of absence, or something like that. . . . We believe that one of our principal reasons for existence is the teaching of undergraduates, and we expect all faculty members to participate in the teaching of undergraduates, and that is a firm requirement. (Interview with Yale administrator)

As a result of the efforts by Yale in general and the OCR in particular, twenty-seven biotechnology companies have been established in the New Haven metropolitan area, and many more are in development, as described in Table 4.2.[9]

Table 4.3 summarizes the accomplishments of the OCR from 1996 until 2012. During that time, annual royalties grew from $5,007,485 in 1996 to $11,023,081 in 2011, a growth of 120 percent. Even when we exclude the impact of the highly successful drug Zerit, we find that Yale's average annual royalty income is about $6.6 million. The university's licenses with income grow to a maximum of $290 million in 2012, with six high-paying licenses.

The average number of US patents issued for Yale University per year grew to twenty-four in 2012. By 2012, in total, Yale had fifty-five spin-outs and twenty-nine biotechnology spin-outs. Figure 4.1 demonstrates the significant increase in Yale's patents between 1983 and 2003, from two to twenty-eight, an increase of 1,300 percent (US Patent and Trademark Office 2003).

Over the past decade, Yale, like many other universities, has created committees to deal with issues such as conflict of interest and the appropriate role of faculty in start-ups. These are the Committee on Conflict of Interest, which creates the policy on acceptable behavior and the need to pursue patenting rights while not limiting research rights, and the Committee on Cooperative Research, which advises the OCR and the provost, who deals with new modes of teaching and learning. Though compared with other universities, Yale is not unique in its creation of such

TABLE 4.2 OCR activities, 1996–2012

| Fiscal year ended June 30 | US patents issued | Licenses w/ income | Annual royalties | Annual royalties w/o Zerit | Start-ups |
|---|---|---|---|---|---|
| 1996 | 12 | — | $5,007,485 | $2,027,606 | 3 |
| 1997 | 23 | — | $13,091,174 | $1,791,128 | 1 |
| 1998 | 25 | 83 | $32,886,208 | $3,234,854 | 6 |
| 2001 | 32 | 119 | $119,500,903 | $4,771,855 | 9 |
| 2003 | 29 | 188 | $11,901,894 | $4,728,067 | 2 |
| 2004 | 27 | 190 | $10,531,084 | $4,592,444 | 4 |
| 2005 | 24 | 212 | $14,209,934 | $8,908,259 | 1 |
| 2006 | 34 | 190 | $20,541,078 | $16,807,099 | 4 |
| 2007 | 18 | 204 | $9,121,960 | $6,487,825 | 6 |
| 2008 | 17 | 213 | $7,767,792 | $5,993,266 | 3 |
| 2009 | 26 | 247 | $8,476,173 | $7,743,674 | 5 |
| 2010 | 25 | 258 | $7,924,033 | $7,803,805 | 5 |
| 2011 | 37 | 207 | $11,023,081 | $10,958,058 | 4 |
| 2012 | 24 | -33 | $5,949,400 | — | 2 |

SOURCE: Office of Cooperative Research (2012).
NOTE: The information provided in this table reflects all fields of technology, not just biotechnology. The information was reported directly to the author and appears only in internal reports.

TABLE 4.3 Yale University, highest-paying licenses in life sciences and biotechnology

| Description | Type | Income amount (millions) |
|---|---|---|
| D4T–Zerit for HIV | Drug | $276.7 |
| L-FMAU–Clevudine for HBV | Drug | $4.2 |
| FTC–Emtriva for Hepatitis B | Drug | $9.5 |
| Cloned Human Tissue Factor | Diagnostic | $3.1 |
| Lyme Rx Vaccine for Lyme disease | Vaccine | $4.3 |
| Diagnosis of Genetic and Malignant Diseases | Diagnostic | $2.0 |

SOURCE: OCR (2012).
NOTE: The university's licenses with income grow to a maximum of $290 million in 2012.

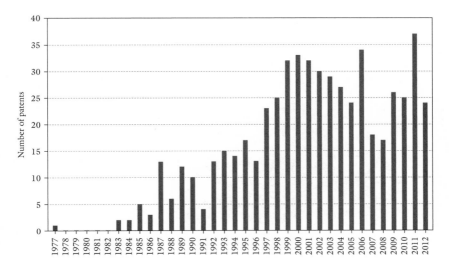

FIGURE 4.1  Yale by patents, 1977–2012
SOURCE:  US Patent and Trademark Office (2012); OCR (2012).
NOTE:  This figure demonstrates the significant increase in Yale's patents between 1983 and 2003.

committees, the committees strengthen the notion that Yale is committed to technology commercialization.

Moreover, all intellectual patents from the inventions of faculty or students belong to Yale. The OCR handles the patenting process. Royalties are on a sliding scale. The first $100,000 of net royalties is divided at 50 percent to the inventor and 50 percent to the university. Net royalties between $100,000 and $200,000 are divided at 40 percent to the inventor and 60 percent to the university. Net royalties exceeding $200,000 are divided at 30 percent to the inventor and 70 percent to the university. This policy results in a high number of university spin-outs. The impact of the policy on university spin-outs and licenses supports existing literature that claims that a lower share of royalties has a direct inverse correlation with the number of spin-out companies (Di Gregorio and Shane 2003). Hence, by allocating a smaller share of the royalties from licenses to the inventor, faculty at Yale are encouraged to spin out companies rather than licensing the technology.

Figure 4.2 provides an example of a Yale royalty deal. The figure, which is relevant to any patent process at Yale, reflects the commercialization of Zerit to Bristol-Myers Squibb.

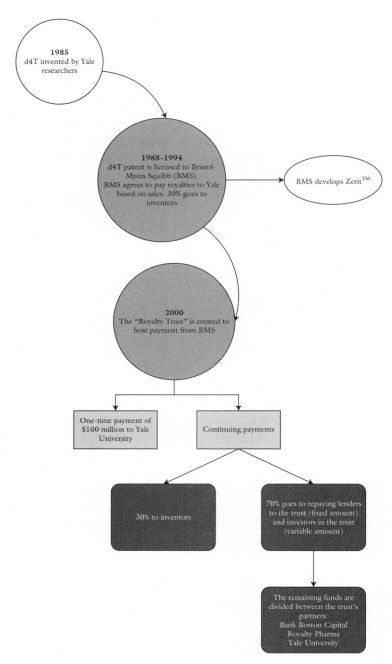

**FIGURE 4.2** How a Yale royalty deal works
SOURCE: Blumenstyk (2001).
NOTE: The process from idea to commercialization at Yale.

## THE IMPACT OF YALE'S ECONOMIC DEVELOPMENT INITIATIVE

It took six years (from 1993 to 1998) to implement the policy, organization, and cultural changes at Yale in general and at the OCR specifically. While this period seems long, the changes at Yale were done with one mission and a specific and constant agenda regarding the ability of Yale to make an economic contribution to its region. The changes at the university had a direct impact on the New Haven metropolitan area and the state of Connecticut, as well as the biotechnology and pharmaceutical companies.

Map 4.1 presents the biotechnology industry in New Haven in 2004. As the map shows, the industry is concentrated in the city of New Haven, divided between the incubator in George Street and the New Haven Science Park, and in East Haven. The few companies we see on the far left next to the New York border are the subsidy of the German pharmaceutical Boehringer Ingelheim and a few companies that formed around it. As will be explained, Boehringer Ingelheim sees itself as part of the New Haven cluster and has created formal relationships with Yale University.

In Figure 4.3 we can see that 59 percent of Yale's biotechnology spin-outs chose to locate in the region. This suggests that Yale had a direct

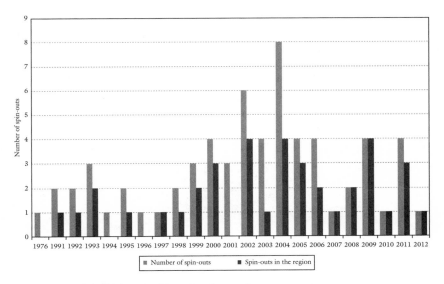

FIGURE 4.3  Yale University biotechnology spin-outs, 1976–2011
SOURCE: Author; Thomson Financials (2011).
NOTE: Fifty-nine percent of Yale's biotechnology spin-outs chose to locate in the region.

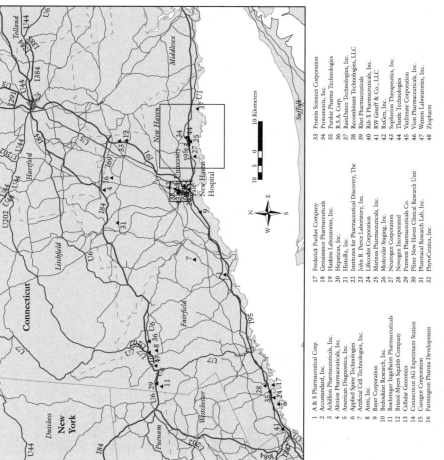

MAP 4.1  The New Haven biotechnology cluster, 2004

NOTE:  Detailed maps of New Haven and East Haven are on the following page. The biotechnology industry is concentrated in the city of New Haven, divided between the incubator in George Street and the New Haven Science Park.

1  A & S Pharmaceutical Corp.
2  Accustandard, Inc.
3  Achillion Pharmaceuticals, Inc.
4  Alexion Pharmaceuticals, Inc.
5  American Diagnostica, Inc.
6  Applied Spine Technologies
7  Artificial Cell Technologies, Inc.
8  Atmi, Inc.
9  Bayer Corporation
10  Bedoukian Research, Inc.
11  Boehringer Ingelheim Pharmaceuticals
12  Bristol-Myers Squibb Company
13  Cellular Genomics
14  Connecticut AG Experiment Station
15  Curagen Corporation
16  Farmington Pharma Development

17  Frederick Purdue Company
18  Genaissance Pharmaceuticals
19  Haskins Laboratories, Inc.
20  Hepaticus, Inc.
21  HistoRx, Inc.
22  Institutes for Pharmaceutical Discovery, The
23  John B. Pierce Laboratory, Inc.
24  Lifecodes Corporation
25  Marinus Pharmaceuticals, Inc.
26  Molecular Staging, Inc.
27  Neurogen Corporation
28  Novogen Incorporated
29  Penwest Pharmaceuticals Co.
30  Pfizer New Haven Clinical Research Unit
31  Pharmacal Research Lab, Inc.
32  PhytoCeutica, Inc.

33  Protein Sciences Corporation
34  Protometrix, Inc.
35  Purdue Pharma Technologies
36  R.S.A. Corp.
37  RainDance Technologies, Inc.
38  Recombinant Technologies, LLC
39  Rhei Pharmaceuticals
40  Rib-X Pharmaceuticals, Inc.
41  RW Greeff & Co., LLC
42  RxGen, Inc.
43  Sopherion Therapeutics, Inc.
44  Thistle Technologies
45  VaxInnate Corporation
46  Vion Pharmaceuticals, Inc.
47  Watson Laboratories, Inc.
48  Ziopharm

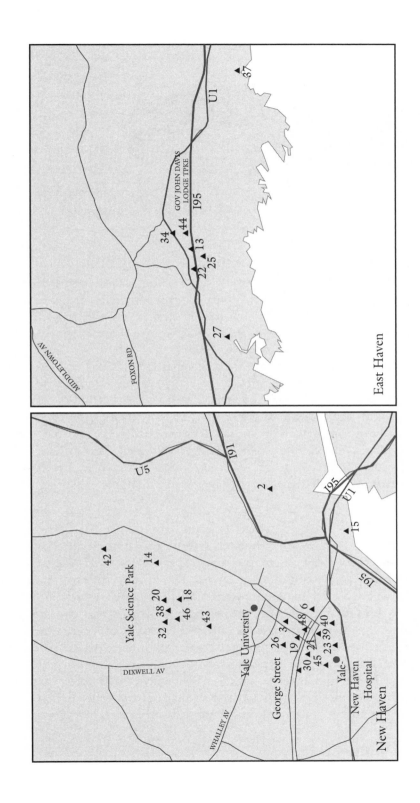

East Haven

New Haven

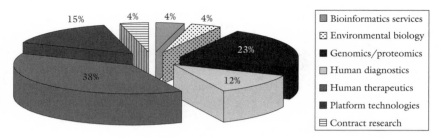

FIGURE 4.4  The biotechnology cluster in the New Haven metropolitan area by sector
NOTE: A majority of firms specialize in human therapeutics.

impact on the location choices made by spin-outs. The wish to stay close to the university implies that companies view university research and resources positively. These findings strengthen the university spin-out literature regarding the importance of locating close to the university of origin (Jaffe, Trajtenberg, and Henderson 1993).

The majority of the biotechnology companies in the area work in the human therapeutic sector. Figure 4.4 shows New Haven companies' strength in biomedicine. This is directly related to Yale's strength in life sciences. The figure includes companies working in more than one sector. The results, below, are drawn from the self-definition of fifteen companies within the cluster. The cluster employs eighteen thousand people directly, and many more indirectly (Connecticut Economic Development 2013). Most of the firms are small to medium size, with fewer than fifty employees.

### Impact on the State of Connecticut

Yale's change toward technology transfer and commercialization made an impact for Connecticut as a whole. Connecticut R&D expenditures in bioscience are constantly growing. The majority of growth can be seen in the biotechnology companies. This has a direct correlation to the growth in total number of biotechnology companies.[10] In 2003, expenditures by the pharmaceutical industry in Connecticut, which dominates the expenses of R&D in the state, accounted for more than 12 percent of all R&D dollars spent by pharmaceutical companies nationwide (CURE 2003). In addition, by 2008, with $4.5 billion invested in R&D, Connecticut ranked third nationally in the percentage of bioscience related R&D expenditures (Rappa 2011). This represents extensive growth compared with the 6 percent spent by pharmaceutical companies in Connecticut in 1995, when there were only six biotechnology companies in the region.

Yale's contribution to the local economy is exemplified by the comment here of a Yale professor who also founded a biotechnology company. In this case, Yale's contribution can be seen in the number of employees purchasing houses in New Haven:[11]

> When I went in to get a new mortgage, I indicated some consulting income from company X. The bank officer stopped and he said, "You're with company X?" and I said, "Yes." He shook my hand, and he said, "You know, you have no idea how many people [from company X] have come to buy houses." That was my first indication that this is how the economy works. My company has an incredible impact. (Interview with Yale faculty member)

Importantly, the university and the state collaborated in the case of the biotechnology industry. Through the Office of New Haven and State Affairs, many issues and problems, such as tax incentives and streetlights, were resolved. This is an additional indication of the inclusion of the technology-transfer process. Yale initiated changes within the university but followed and collaborated with the state and the city to achieve the desired economic goal.

What is interesting here is that, although Yale has been the catalyst behind the formation of many of the biotechnology companies in the cluster, it is not the leader of the cluster and does not view itself as such. In particular, Levin and the university saw themselves as drivers for change. They wanted the university to change and make an economic contribution to the region with, first and foremost, a benefit for the university itself by becoming a safer school that could compete for the top minds in the fields. The university did not invest its own funding or properties in the development of the biotechnology cluster. From interviews with company chief executive officers and organizational leaders, it became clear that several local organizations, including CURE (industry association) and the Connecticut Office of Bioscience, were attempting to perform this role in several ways.

Perhaps because there has been no clear leadership within the cluster, the local biotech agglomeration in New Haven has not functioned as a classic industrial district. According to interviewees, one of the reasons for the inability of the cluster's companies to collaborate and communicate with one another has been the divided leadership in the cluster. Although the cluster grew significantly in the past decade, it has not yet developed

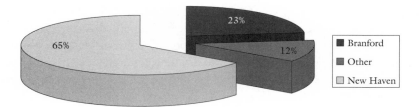

FIGURE 4.5   New Haven biotechnology companies by location
NOTE: The companies are spread among three locations.

the communication and networking relationships among companies, which are indicators of a successful cluster.

Furthermore, as can be seen in Figure 4.5, the companies are spread among three main locations that are at least fifteen to twenty minutes (by car) away from each other, and only the George Street facility has an unofficial area for meetings nearby (e.g., coffee shops, restaurants). The lack of meeting space has created a situation in which none of the smaller clusters has enough companies to create the institutional thickness needed for networking (seen in other clusters in other regions) through unofficial meetings and the movement of employees among firms in the cluster. The official opportunities for interaction occur through Yale's seminars and conferences, as well as meetings convened by CURE and other organizations, such as PriceWaterhouseCoopers or Connecticut Innovations.[12] However, these meetings are known only to part of the biotechnology community and do not provide a constant and regular way of exchanging information.

### The Impact on the Local Pharmaceutical Companies

Yale's cultural change toward disseminating academic ideas to the private market resulted in a larger regional change in the culture of technology transfer. The strength of the local biotech industry has changed the way existing pharmaceutical firms interact with other players in the region. The local pharmaceutical companies have significantly changed their behavior and funding patterns, giving more attention to the local intellectual base. There are ongoing collaborations between local pharmaceutical firms and the local universities and research institutes, cultivated by Yale's OCR, CURE, and the Office of Bioscience. These connections include, but are not limited to, Yale and the University of Connecticut. Pfizer, for example, chose to utilize the local knowledge base by developing a direct relationship with Yale. Pfizer invested $35 million in a sixty-thousand-square-foot clinical trial facility in downtown New Haven between Park and Howe Streets, on land owned by the State of Connecticut. The importance of this facility is that it is a shared research facility between Yale and Pfizer. Although some parts of the projects are confidential, patients are moved between the company and the university. Additionally, Bayer initiated a scholar's program in 2003 that appoints a faculty member each year as a fellow who works closely with Bayer.[13]

Since the late 1990s, one can find business relationships between local pharmaceutical firms and the local biotechnology industry. Neurogen Corporation, a biotechnology company, and Pfizer began a two-year research partnership in 1998 to work on GABA neurotransmitter receptor-based drug programs for the treatment of anxiety, sleep disorders, and cognition enhancement. Bayer and CuraGen Corporation collaborate on obesity and diabetes in codevelopment, pharmacogenomics, and toxicogenomics.

## The Impact on the Biotechnology Industry

A university's contribution to a cluster is determined not only by the academic strength of the university but also by its ability to contribute to the cluster through networking, services, and support, as well as by the receptivity of the region (Di Gregorio and Shane 2003; Etzkowitz et al. 2000; Feldman and Desrochers 2003; Keeble 2001; Lawson and Lorenz 1999; Morgan 1997). Yale's changes toward technology commercialization have affected the way industry views the university and responds to the university's academic contribution; the university's active involvement in the development of new companies; its role as a facilitator between companies, and its assistance to companies' access to resources, such as venture capital, management, and science parks.

Aside from the fact that many of the local biotechnology companies know they would not have been created without Yale and the Office of Cooperative Research, they view the OCR in particular and Yale in general as significant players in the development of the local biotechnology cluster. An executive of one biotechnology company stated that Yale is "absolutely critical. Yale is the main driver of further development of the cluster." Since there were only a few companies prior to Yale's initiative, the OCR had to assist companies in their development. Inside the university, the OCR recruited business students to assist university spin-outs in the writing of a business plan. Yale's participation has been crucial, as is evident from the remarks of one of the companies' executives:

> [Yale is] very important. They play a very important champion and facilitating role, [in] formulation of business plans and also sponsoring companies while they are raising capital. Companies may choose not to use it but it is there. There is direct involvement of several people at Yale: Jon Soderstrom and Greg Gardiner in the choosing of management, raising capital, writing business plans. (Interview with biotechnology executive)

As with the recruitment of star scientists and management for the university itself, Yale's administrators use their contacts and the Yale name to connect their companies with venture capital firms. Fortunately, a community of venture capitalists exists in the area of Stamford and Greenwich, Connecticut, a short drive from New Haven. However, the OCR had to convince venture capital firms that investing in companies that choose to stay in the region is very important for both Connecticut and Yale.

Another important point mentioned by several companies is Yale's involvement with the creation of science parks. According to one biotechnology company executive, "There would be no Science Park without Yale. Yale is one of the last Ivy Leagues that see the importance of fostering technology. Yale is the intellectual and economic driver."

By recruiting Bruce Alexander and creating Yale's Office of New Haven and State Affairs, Yale took it upon itself to promote the availability of offices and laboratories. Interviews with executives at companies created before the development of science parks revealed that the lack of laboratory space caused companies to divide their activities among several floors of a building, which created a situation in which one part of the lab was on the first floor and the other was on the third floor.

Incubator space not available in the past now exists in the George Street building, next to other biotechnology companies and between Yale's main campus and its medical school. This proximity allows young companies to stay in contact with their founders, who are mostly faculty at the medical school:

> It [the incubator on George Street] is in close proximity to our founders' lab. That enables us to have regular contact with the founders and the people in the lab and get consulting from them. It is close to the school of medicine, which allows us to get access to resources like the school's library. The building is particularly attractive because it is a biotech building; there are a number of biotech companies there, which allows us to borrow equipment. (Interview with executive at a biotechnology company in the incubator)

Other ways in which Yale's policy change influenced local companies is evident in Yale's open-door policy. With this policy, companies participate in seminars and use the Yale library and equipment. However, as an academic institution, Yale gives priority to its students and faculty; companies can use those resources, but Yale does not publicize their availability. Additionally, many of the companies' employees graduated from Yale, have joint research projects with Yale faculty, recruit employees and interns from Yale, and conduct their clinical trials at Yale–New Haven Hospital:

> [Yale University is] very important: talent, the training of scientists. Faculty and scientists are advisers and consultants [to biotechnology companies]. You can do clinical trials [at the hospital]. The proximity to Yale helps recruit people to the area. (Interview with biotechnology executive)

Many interviewees expressed their appreciation for the changes in policy and attitude taken by Yale and the OCR, and believe that without the "push" from the university there would not have been a viable biotechnology cluster in New Haven today.

SUMMARY

University technology transfer is important to local economic development. Changes to technology transfer at a university can positively affect its ability to commercialize technology and interact with the local industry. This chapter illuminated the finding that many factors need to be considered when universities engage in local economic development; the intensity of the change within the university is important. Yale's changes were not limited to its technology-transfer office but were institution-wide and covered policy, organization, and cultural changes. Also, the inclusion of other regional players matters. Yale did not operate or create the biotechnology cluster on its own. The university collaborated with the state, the city, and local industry to push for local economic impact. Finally, the velocity of change affects the ability of the university to adjust. The changes at Yale University started in 1993 and ended in 1998, a relatively short and concise move toward technology commercialization and local economic development. No additional changes have happened since that time.

In summary, the changes Yale implemented were done in a steady, calculated way, considering leadership and areas of expertise, and hence placing importance on the intensity, inclusion, and velocity of the change. By conducting a complete institutional change, Yale University went through an intensive internal change that included not just the technology-transfer office but also the Office of New Haven and State Affairs and the general attitude toward the role the university plays in the region. Second, the change was done in one process that took five years in which all of the possible aspects of the change were done. Last, we find the success in Yale's change through its inclusion of other regional players, which created a cooperative regional change that led to local economic development.

The next chapter analyzes the role played by the University of Cambridge in the creation and development of the local biotechnology cluster in Cambridgeshire, in the United Kingdom. In contrast to the changes at Yale University, technology-transfer changes at Cambridge had a negative impact on the university's ability to commercialize technology and on the local economy. The university made many of the changes due to outside pressure, without a long-term plan, which resulted in several periods of localized changes and did not include other regional players.

## CHAPTER FIVE

# The University of Cambridge

The University of Cambridge ranks as the seventh-best university in the world and the second best in biomedicine research (*Times Higher Education* 2013). Moreover, with 154 companies in 2013, Cambridgeshire County is the largest biotechnology cluster in the United Kingdom, which, after Germany, has the largest number of biotechnology companies in Europe (Dun and Bradstreet 2005; Eastern Region Biotechnology Initiative 2005–6).

The region is famous for its high-tech cluster, which become known as the "Cambridge Phenomenon" after the publication of a Segal Quince Wicksteed report, and for its university-industry relationships (Segal Quince Wicksteed 1985, 2000). The region's environment is rich with academic excellence in the biomedical sciences, with world-class institutions located in close proximity to London. Moreover, private stakeholders in the form of Trinity College, St. John's College, and Barclays Bank, which rely on the excellent research base in the region, contributed to the creation of this vibrant biotechnology cluster. At Cambridge, it is presumed that faculty will be involved in consulting and research with industry.

This biotechnology cluster is a result of a strong university-industry relationship in the region in general, with Cambridge University as a central player. However, in the late 1990s Cambridge started a series of technology commercialization policy and organizational changes. The changes were spread over a period of eight years, were done only at main technology-transfer offices, and were not done in collaboration with any other university or regional players. The changes ended up severely damaging university-industry relationships and the ability of Cambridge to commercialize technology. This chapter follows those changes and their impacts.

Although many of the companies in Cambridgeshire County rely on technology invented at the university, this book argues that the University of Cambridge, though important, did not play an active role in the creation and the development of the biotechnology cluster. Moreover, it was changes to the original technology commercialization system that damaged the university's ability to commercialize technology.

## EXTERNAL UNIVERSITY FACTORS

The University of Cambridge is in Cambridgeshire County in the east of England, forty-four miles northeast of London, forty-five minutes away by train. Cambridgeshire itself has a small airport that provides direct flights to Europe. The county's population in 2011 was 612,590, of which 120,910 were residents of the city of Cambridge.

The system of innovation in the region consists of universities, research institutes, hospitals, local and national government representatives, and private industry. There are six colleges (Cambridge Regional College, Huntington Regional College, College of West Anglia, Isle College, Long Road Sixth Form College, and Hills Road Sixth Form College) and two universities in the county (Anglia Ruskin University, formerly known as Anglia Polytechnic University, and the University of Cambridge). Like Oxford University, Cambridge University has a unique structure and organization. Composed of colleges and departments, the university is decentralized, with each of the colleges responsible for its own finances, policy, and student recruitment. Moreover, central administration at Cambridge was very weak to nonexistent until the 1990s.

Apart from the University of Cambridge, Cambridgeshire has several research institutes focused on biomedicine, which together provide the local system of innovation for the biotechnology industry. The Babraham Institute is an educational charity devoted to biomedical research sponsored through the Biotechnology and Biological Sciences Research Council. The Sanger Centre is one of the centers involved in sequencing the genome. The European Bioinformatics Institute, which moved to Cambridge from Germany, is an academic organization, part of the European Molecular Biology Laboratory and a center for research and services in bioinformatics. The Laboratory of Molecular Biology is part of the Medical Research Council and Addenbrooke's Hospital, the teaching and research hospital for the University of Cambridge. Moreover, the region is home to many local, national, and international high-technology and biotechnology firms (Segal Quince Wicksteed 2000).

Several reports and local policies influenced the region's development over the years. After World War II, local government agencies and the university formed a committee to focus on local "problems" caused by postwar growth. Professor William Holford from University College, London, headed the committee. The report by the Holford Committee, published in 1950, concluded that all growth of buildings, population, and businesses should be limited. The conclusion was based on the belief that the local population did not want further economic development, believing that the city's historical value would be compromised. In fact, the report did not consider any possible benefits that could stem from business growth in the region and thus contributed to the antidevelopment environment in which the University of Cambridge operated. However, the Holford Report encouraged the expansion of research activities, specifically those with connections to the university. The conclusion of the report was soon tested by the rejection of IBM's request to open an R&D facility in Cambridge. The company's blueprints were refused on the grounds that "research laboratories employing a thousand or more scientists and technicians would be undesirable for this area." Thus, IBM chose to open its facility and provide employment for "thousands or more scientists and technicians" in Switzerland instead (Koepp 2002, 192; see also Segal Quince Wicksteed 1985; Garnsey and Hefferman 2005).

Seventeen years later, in 1967, because of pressure from faculty at the university and local research institutes, the university established a subcommittee to examine the relationships between university and industry. Professor Neville Mott, director of the Cavendish Laboratory at Cambridge, headed the committee. In 1971, Cambridgeshire County adopted the report, published in 1969, which supported the expansion of research facilities in the area. The report recognized that the Holford Report had created difficulties in recruiting university faculty and restricted the growth of industry in the region. Even though the report recommended the allocation of small-scale commercial research activity and the creation of a university science park, at the same time it placed limitations on large-scale manufacturing facilities (Committee on Higher Education Under the Chairmanship of Professor Nevill Mott 1967; Segal Quince Wicksteed 1985; Garnsey and Hefferman 2005). Thus, the Mott Report was an important factor in the development of university-industry relationships in the region.

One result of the Mott Report was the creation of the Cambridge Science Park in 1970 by Trinity College (Garnsey and Hefferman 2005).[1] This was a real estate investment by the college, which for many years, and because of its high rents, was mostly a home for international and large companies. It was only in the early 2000s that the Science Park changed its policy and provided smaller units with shorter rent periods, allowing small and medium-size enterprises to locate in the park (Gray and Damery 2003; Federation of Finnish Technology Industries 2010). Importantly, the Science Park is not located near the university or city center. It is not easily accessible by bike, and there are no transportation connections between the park and other science parks south of Cambridge, the train station to London, or Addenbrooke's Hospital. Hence, the park's location does not promote information flow from university labs, students, and so on that can benefit resident companies.

In 1987, another science park, the St. John Innovation Centre, was created by St. John's College, catering mainly to high-tech companies; it has very little laboratory space and hence is not suitable for biotechnology firms. As in the case of the Cambridge Science Park, the center was created to maximize the college's return on its investment by providing shared administrative and office space to smaller companies. Since its creation the park had three directors. Bill Bolton was the director for the first three years, followed by Walter Herriot, who managed the center until his retirement in 1998.[2] Herriot is an important figure in the history of the Cambridge high-tech cluster. Starting in 1978, while Herriot was director of the local branch of the Barclays Bank, the bank supported new ventures in the Cambridge area. This initiative, and especially the virtual office space provided by the park, is seen as instrumental to the growth of the high-tech cluster in Cambridge (Segal Quince Wicksteed 1985). The creation of science parks by the two colleges might appear to be a university initiative; however, because of the unique structure and independence of the colleges, these actions must be understood as separate acts by each college.[3]

Historically, the region has been influenced by legislative factors in the form of national government initiatives to support local industry. These efforts are channeled through the British Department of Trade and Industry (DTI).[4] The DTI was represented in the region by the East of England Development Agency (EEDA), one of nine regional development agencies created in 1999 to transform England's regions through sustainable economic development. Its task was "to improve the region's economic performance and ensure the East of England remains one of the UK's top performing regions" (EEDA 2005). It is important to note that EEDA was responsible for redistributing the wealth and growth in the region. Moreover, the UK government, even though proud of such thriving regions as Cambridge, focuses its infrastructure support on the poorest regions of the country. This results in underinvestment in the Cambridge region (Gray and Damery 2003). The EEDA's agenda in life sciences was to understand the needs of the industry, such as the shortage of technicians and entrepreneurs, and to stimulate debate. However, unlike the United States, the UK agency did not provide direct funding, or tax incentives, for the companies, but it did provide £82 million ($11.5 million) in support of different programs and institutions, such as One Nucleus, the industry association discussed later in this chapter, as well as the association's annual conference and other programs, such as i10 and Babraham BioConcepts.[5] In March 2012, EEDA and a few other regional agencies were terminated. The Local Enterprise Partnership took its place. One of its roles is providing funding to invest in business assistance, skills, and research commercialization (Greater Cambridge Greater Peterborough Enterprise Partnership 2013).

In August 2003, DTI launched a venture capital fund in each region. In the east of England there is a £20 million fund managed by Create Partners. The fund invested in small and medium-size enterprises that can demonstrate significant growth potential and that are based in the east of England. Relatively small venture capital investments of up to £250,000 or $360,000 initially are available, with the potential for further investments of up to £250,000. However, the funds are not directed at university spin-outs or biotechnology (EEDA 2005).

Another program and an important historical factor that benefits the biotechnology industry indirectly is DTI's Technology Program. Over the period 2005–8, £320 million was available for industry in the form of R&D grants in the technology areas identified by the technology strategy board. Financial support of university spin-outs was available for the University of Cambridge and the Babraham Institute through the £3 million Challenge Fund; however, the funds were available for all spin-outs, not just biotechnology. An examination of private venture capital firms in the region finds that Cambridge, even though a short distance from London, has several venture capital firms in the city (Cooke 2002). Moreover, London has the largest concentration of venture capital firms outside the United States.[6]

Besides the government initiatives, industry created its own institution to promote life sciences in Cambridgeshire. Since 1997, the Eastern Region

Biotechnology Initiative, now known as One Nucleus, has represented the biotechnology industry in the region. The group includes representatives from the local biotechnology industry, pharmaceutical companies, research institutes, and government organizations such as EEDA. One Nucleus offers services in the form of networking and special interest groups, providing information on finance, human resources, and business development.

Cambridgeshire County does not offer specific incentives to the biotechnology industry. However, it benefits from incentives created by the UK government. These can be seen in the form of the Bio-Wise program, a £14.5 million ($20.8 million), five-year DTI program that ended in 2004, aimed to help UK manufacturing companies increase their competitiveness through the use of biotechnology in their products or processes.

In summary, Cambridgeshire County's biotechnology and university-industry relationships were influenced by both historical and environmental factors. Neither the City of Cambridge nor the County of Cambridgeshire has favorable policies or funding structures to support business creation. The work of the Holford Committee and the Mott Report created regional zoning that favors only research facilities and small businesses. Furthermore, local government funding is scarce. The regional agency EEDA did not provide any financial support for companies, as in general it supported poorer areas of the county. Other government programs, such as the University Challenge Fund, were short term.

INTERNAL UNIVERSITY FACTORS

Most studies of the Cambridge economy, which focused on the university's academic excellence and regional social networks, argue that the university was a fundamental driver of growth (Garnsey and Hefferman 2005; Segal Quince Wicksteed 1985, 2000). These studies portray the unique capabilities of the university and its faculty and students in commercializing technology, licensing, patenting, and spinning out companies. As the next section depicts, this spirit of entrepreneurship existed in the university as early as the 1880s.

As a result of governmental pressure and resource availability in the late 1990s, the university began a series of policy and organizational changes to promote technology commercialization. These efforts were done through the central administration of the university, with no coordination with the colleges, industry, and other regional players. Moreover, the efforts ignored a long and successful history and culture of innovation and entrepreneurship, and hence disturbed a historically successful process of technology commercialization.

## Culture

This book strengthens existing studies that praise the Cambridge entrepreneurial culture. As this section shows, from its inception, Cambridge University has been a leader in research, innovation, and technology commercialization. Cambridgeshire County has been called the "Cambridge Phenomenon" and "Silicon Fen" for its successful development of the IT industry (Koepp 2002; Segal Quince Wicksteed 1985, 2000). Many of its IT and now biotechnology firms have direct roots in the university.

It is helpful to examine the university's mission statement in order to understand the importance of economic development to the university. O'Shea and colleagues (2005) argue that a university's history and resources influence its mission statement and organization, and also influence university technology-transfer capabilities. The organization's cultural base, influenced by the organization's history and the history of the decision makers in the organization, affects the way in which the organization makes decisions about issues such as strategy, outlook, and cooperation with other players in the local economy (Schoenberger 1997). The University of Cambridge sees itself as a national and international university, not as a leader in the region where it is located. This self-representation influences the way the university engages and contributes to the local economy. According to the university's mission statement: "The mission of the University of Cambridge is to contribute to society through the pursuit of education, learning, and research at the highest international levels of excellence" (University of Cambridge 2005b). This mission statement also reflects the way in which the university structures university-industry relationships in general and technology transfer in particular.

The original structure of the university has an impact on its commercialization and entrepreneurial culture. The University of Cambridge was established in 1209 by scholars who left Oxford. The first college, Peterhouse, was established in 1284, creating the foundation for the collegiate system—a defining feature of the university. The colleges are autonomous institutes that select their own faculty and students although they are connected to the university through membership in the university council and representation on the different boards. Thus, college faculty do not have

positions in a university department, and faculty in a university department do not have to hold a position at a college. Students, in contrast,
always belong to both the university and a college. The colleges differ by
financial capability, educational strengths, and students. Three accept only
women, and most accept both undergraduates and graduate students. This
diverse and disperse system of departments, colleges, and central administration affects the commercialization culture of the university. As with
the spirit of college independence, several colleges and several departments
created their own technology-transfer offices or programs that work in
parallel to the central administration ones.

The Cambridge University entrepreneurship culture is evident as early
as the 1880s in the development of the Cambridge Scientific Instruments
(CSI) and the Pye Company. Horace Darwin created CSI in 1881 to design
and manufacture equipment for the University of Cambridge. Darwin, the
son of Charles Darwin and a member of the university, maintained a close
relationship between his company's research and the university. In 1896,
W. G. Pye, a mechanic in the physics laboratory at Cambridge, created a
company to design and produce laboratory equipment. It would become
a pioneer in radio and television equipment. As in the case of CSI, the
Pye Company worked closely with the university (Segal Quince Wicksteed
1985). Led by CSI and Pye, Cambridge was a hub for technology-oriented
industries throughout the twentieth century (Koepp 2002; Garnsey and
Hefferman 2005).

Though both firms represent the early innovation and entrepreneurship at Cambridge, they also represent the local business culture, in which
a five-hundred-employee company is considered large (compared with the
United States, where five hundred employees represent a medium-size
firm). Over the years, both companies showed an unwillingness to seize
their growth potential. The Pye Company, for example, chose to sell its
radio business to C. O. Stanley, the company's advertising agent, who later
bought the mother company to create the Pye Group, which became one
of the biggest employers in the Cambridge area. In the mid-1960s, due to
financial problems, the Pye Group was bought by Phillips. Despite financial difficulties, CSI chose to stay dedicated to university needs and did
not search for new business opportunities. This decision placed the company in a difficult position, and in 1960, after failed attempts to market its
microscope on a worldwide basis, the company could not hold up against
industry pressures and was forced to merge with George Kent Ltd. (an
engineering company). The UK government pressured CSI to merge with

Kent through the purchase of company shares with £4 million ($5.7 million) of the public's money.[7] These two leading Cambridge companies are examples of successful science- and technology-based companies, whose choice to stay small led to being either bought out or merged with larger international corporations.

The strength of individual entrepreneurs in the Cambridge cluster is a well-studied phenomenon (Garnsey and Hefferman 2005; Myint, Vyakarnam, and New 2005). According to those studies, the base of Cambridge's development is social capital. Furthermore, the success of companies in the cluster is based on a small number of connections among entrepreneurs. Although many of the entrepreneurs have some connections to the University of Cambridge, becoming part of what Myint, Vyakarnam, and New (2005) refer to as the "mini-clusters" is a more significant factor leading to the success of a company. Myint, Vyakarnam, and New (2005) claim that within the Cambridge cluster, it is possible to observe several mini-clusters of companies associated with key entrepreneurs. Hence, the University of Cambridge has been promoting innovations and entrepreneurship in general research from its inception. However, this spirit was a general one that was not focused solely on the region in which the university resides.

### Policy

In its early days, as a result of lack of funding and a policy in which faculty owned their own inventions, Cambridge University had a relaxed policy toward technology commercialization. Starting in the late 1990s, following governmental pressure, the university went through a transition from a technology-transfer approach in which the faculty had what is known in Cambridge as a "free hand" to the current model, characterized by its centralized control mechanisms.

Intellectual Property Rights Policy

Intellectual property rights (IPR) policy in Cambridge has changed dramatically since the 1960s, with the latest change implemented in December 2005 (*Cambridge University Reporter* 2005). For many years, the University of Cambridge was known for its relaxed, less controlling "policy toward commercial exploitation of academic know-how and links with industry generally" (Segal Quince Wicksteed 1985, 47). In this quote, the authors of *The Cambridge Phenomenon* were referring to the free hand academics were given in Cambridge in terms of research commercialization.

Until 1998, intellectual property rights in Cambridge had a flexible framework. Officially, if they were supported by a research grant, faculty had to consult with Wolfson Industrial Liaison Office (WILO), the university's technology-transfer office. However, the intellectual property was not automatically assigned to the university, and in many cases, the research sponsors or the faculty themselves took the ownership of the invention (Minshall, Druilhe, and Probert 2004). This policy was based partly on the fact that the university did not have the resources to take a leading role in the matter and also depended on the individual goodwill of the inventors to provide the university with a share in future royalties (Segal Quince Wicksteed 1985). This flexibility was unique in comparison to other UK universities. All of the other universities applied the 1997 Patent Act, by which the university owns the rights of inventions based on sponsored research. The responsibility of WILO was to provide the outside world with an access point to the university, as well as assistance to industry and academic faculty in contractual, patenting, or licensing issues. As mentioned by Richard Jennings (1994, 61), while he was director of WILO: "There is a general recognition of the important strategic value of relationships with industry and a commitment to their realization but this is approached largely by assuming that the demonstration of the excellence of its research result rather than by an emphasis on excessively formal policies and centralized bureaucratic systems."

This laid-back, flexible approach to intellectual property rights changed in 1998–99, when Cambridge began to take a more structured approach to managing its employees' inventions. This was the result of governmental pressure and funding to promote technology transfer (Hatakenaka 2002; Statistics Finland 2010). These funds were the basis of the change in the technology-transfer offices at Cambridge and the creation of the Research Services Division, which are described in the following section. Although the 2001 policy made no changes to the official definition of intellectual property ownership rights, it did clarify existing rules. As cited in the *Cambridge University Reporter* (2001):[8]

> IPR generated by Externally Funded Research, except where the University has agreed otherwise, will be owned by the University;
> The University will not claim ownership of copyright in normal academic forms of publication such as books, articles, lectures, or other similar works generated in the course of Externally Funded Research unless those works have been specifically commissioned by a sponsor;

Where revenue is generated by the exploitation of IPR arising from Externally Funded Research, any net benefit received by the University, after deduction of agreed costs, will be shared between the inventor, his or her Department and the University on the following terms, which shall be revised from time to time.

On December 13, 2005, the latest change in the university's IPR reforms was made (*Cambridge University Reporter* 2005). The university now receives control over all inventions, regardless of the source of funding. According to university officials, this move provides more academic freedom to inventors, as it secures their copyrights and "upholds the right of academics to place their inventions in the public domain" (University of Cambridge 2005a; Statistics Finland 2010). Inventors must go through the University of Cambridge's technology-transfer office, Cambridge Enterprise,[9] which patents the invention and decides whether to license the technology or spin out a company. If Cambridge Enterprise is involved in the commercialization process, income from the invention is distributed among the inventor, the department, and the university (see Table 5.1).

If the invention arises from a nonformal university activity of a university staff member, the inventor can choose to commercialize the invention by him- or herself. In this case, the inventor will carry the financial burden and be obliged to pay a percentage to the university and the department (see Table 5.2). In the case that a spin-out company is based on an invention, the university has equity in that company. The equity percentage varies from case to case.

This last reform in the university's IPR had a greater impact on venture capitalists and biotechnology companies than on university academics. Many feared that the change would prevent and obstruct the technology transfer and commercialization process, and would not allow access to

TABLE 5.1 University of Cambridge's share of revenues from net royalties where Cambridge Enterprise is involved in the exploitation

| Net income | Inventors (jointly) | Department | Cambridge Enterprise |
|---|---|---|---|
| First £100,000 | 90% | 5% | 5% |
| Next £100,000 | 60% | 20% | 20% |
| Above £200,000 | 34% | 33% | 33% |

SOURCE: *Cambridge University Reporter* (2005).
NOTE: Up to £200,000 royalties are in favor of the inventor.

TABLE 5.2   University of Cambridge's share of revenues from net royalties where Cambridge Enterprise is not involved in the exploitation

| Net income | Inventors (jointly) | Department | Central funds |
|---|---|---|---|
| First £50,000 | 100% | 0% | 0% |
| Above £50,000 | 85% | 7.5% | 7.5% |

SOURCE: *Cambridge University Reporter* (2005).
NOTE: Royalties are in favor of the inventor.

university intellectual property. According to Segal, Quince, and Wick-steed (1985), authors of *The Cambridge Phenomenon*, Cambridge's growth in high tech and biotech was based on the earlier free-hand approach. Some of the biotechnology and venture capital executives interviewed for this study already viewed the changes in the university as a problem, believing that the IPR royalty changes will make the licensing process more difficult and bureaucratic. The following quotes represent the negative view on the policy changes projected by local biotechnology executives who in the past had numerous dealings with the university:

> If I can choose to license technology from Cambridge or from Imperial or Oxford, I will choose Imperial or Oxford over Cambridge. (Biotechnology executive)

> I think they [university administrators] are a barrier. They have a very difficult job sitting between ourselves and the academics. In my perspective they are obstructive. The minute we decide to build a company time scale is very important to us. They are very slow. Not on a business scale. The companies are formed in spite of them. (Venture capitalist A)

Believing that changing the IPR ownership from faculty will have negative effects on basic research, many academics rejected the new policy. Moreover, the success of Cambridge companies was attributed to the original faculty-owned IPR and a relaxed commercialization policy (Garnsey and Hefferman 2005). It took three attempts and a vote of 790 out of 1,098 on October 5, 2005, to pass the policy in the Regent House (Statistics Finalnd 2010; *Cambridge University Reporter* 2005). A few academics did not view the change in IPR as a problem, claiming that the change would not harm them and that the university should be acknowledged for its part in the invention's research:

It was important for us to acknowledge the university contribution. The university was given a share [in the company]. (Faculty and company founder C)

Thus, IPR policy at the university has gone through many dislocating changes since the 1960s. In the last change, Cambridge University became the sole owner of technology originating from university research.

Patents

On examination of the patents assigned the University of Cambridge, we find that up to the mid-1990s the university produced about six or seven patents per year, and starting in 1996, in conjunction with the high-tech boom, these figures doubled and reached a peak of ninety-eight patents per year in 2002. Compared with other highly ranked universities, such as MIT, California Institute of Technology, Stanford, and Johns Hopkins, Cambridge is one of the top university patent owners in the world (US Patent and Trademark Office 2003).[10]

When analyzing the growth of university patents, we need to take into account two important factors. First, when we consider the fact that up to 2005 most inventions were owned by individual faculty, and hence not automatically assigned to the university, it is remarkable how well the university did. This means that many faculty still chose to assign the patent to the university. Moreover, the growth starting in 1998–99 indicates that the push by central administration to commercialize technology had a positive impact. That said, the chart is somewhat misleading as to how much growth in patents the university has seen. Faculty have not changed their patenting frequency, but the university now has more patents assigned to its name (see Figure 1.1).

Second, to support the strength of patenting at Cambridge, we need to examine research funding. Though government funding for higher education in the United Kingdom has been growing, the total funding per student and university has declined. However, we see a growth in patents in spite of this decline. Interestingly for Cambridge, although government funding declined from 1999 to 2004, the University of Cambridge's income grew by 128 percent, including an increase of 82 percent in income from grants and contracts. The income continued to grow and by 2011 reached £840,761,000.

Data on the breakdown of the university's income can be seen in Table 5.3. Note that 50 percent of MIT's income is based on federal government funding, whereas only 23 percent of Cambridge's income is based on

TABLE 5.3 University of Cambridge income, 2011–12

| Source | Amount (millions) | Percentage |
|---|---|---|
| Government funding (HEFCE, HEDCE, and the Teachers' Training Agency) | £197.2 | 23% |
| Fee income | £149.2 | 18% |
| Other income | £146.4 | 17% |
| Endowment and investment income | £54.3 | 6% |
| Research grants and contracts | £293.4 | 35% |
| **Total income** | **£840.7** | **100%** |

SOURCE: University of Cambridge (2013).

government funding. Another 35 percent is based on other funding for research grants and contracts, including private industry. About 40 percent of the research grants and contracts income comes from the UK government, which places total government funding at 36 percent.

### Spin-Outs

In 1987, the University of Cambridge started to officially spin out biotechnology companies. By that time the region already had five biotechnology companies. However, until the privatization of the British Technology Group in 1991, most university spin-outs were not accounted for, and intellectual property rights belonged to the inventor and not the university. It was up to the inventor to create the company, with no assistance from or association with the university.

Though companies have been forming based on university research, and officially spinning out of university laboratories, Cambridge University does not have direct policy relating to spin-out companies. Moreover, the number of university spin-outs that originally occurred in the late 1990s, in the beginning of the change period, saw a sharp reduction after 2001, from which the university has not yet recovered (as of 2012).

The university provides a home to many of the university spin-outs, which in many cases start at the inventors' labs. Furthermore, Cambridge Enterprise offers incubator space at the new William Gates computer science building on the west side of campus. However, this is a very limited space, not capable of hosting a wet lab, making it less useful for biotechnology firms.

Cambridge does not aim to keep spin-outs in the region; yet 84 percent of the forty-four biotechnology university spin-out companies chose to stay in the region. This is not surprising, since the current location of a firm has a direct correlation to the first location of the spin-out and its first employees.[11] Information from interviews and from other studies indicates that the community, culture, and lifestyle of the region contribute to retention of skilled people. Thus, when a firm originates from the university, and in many cases the management and employees are recruited from the university as well, the company chooses to locate in close proximity to it (Breznitz 2000; Breznitz and Anderson 2006; Eaton and Bailyn 1999).

Interviews revealed that many of the spin-out companies maintain close relationships with the company founder, who is a faculty member or researcher at the university. Interestingly, and different from the situation at Yale and most US universities, at Cambridge faculty can hold dual positions at the university and at their company. Hence, in many cases, the founder has an official role in the company, varying from the position of chief executive officer to a seat on the company's scientific advisory board. Furthermore, many of the companies have started inside a university laboratory while employing laboratory students and later university graduates. This situation also differs from that at Yale, where the company is generally located in a space other than the faculty member's university lab.

When we examine investment in the biotechnology university spin-outs, we find that firms have succeeded in recruiting professional venture capital funding (see Table 5.4).[12] Interestingly, in comparison to the Yale spin-outs, these figures represent about a third of the venture capital funding recruited by Yale spin-outs. These figures, as indicated by studies on technology firms, may indicate the rate of success for the firms (Kortum and Lerner 2000). Hence, the low level of investment in university spin-outs may indicate the lack of confidence that local venture capital firms hold for these firms and the university's technology-transfer office.

A close examination of the Cambridge biotechnology spin-outs in Figure 5.1 finds a decline in the number of spin-outs starting in the year 2001. The total number of spin-outs grew steadily from one in 1991 to six in 1999, after which there is a sharp decline in 2000 to only one company, growing back to seven spin-outs in 2001 and dropping by one or two companies every year since. This trend is clear even if one takes into account the decline in the number of university spin-outs in 2000, which may be attributed to the high-tech crises of the late 1990s and early 2000s. Rather, and as the following paragraphs will show, this decline was unique and can be explained only by the organizational changes in the university's technology-transfer office (see Figure 5.2 later in the chapter).

TABLE 5.4  Cambridge biotechnology companies by venture capital funding, 1996–2012

| Name | Year founded | Location | Total VC funding |
| --- | --- | --- | --- |
| Mission Therapeutics | 1996 | Cambridge | $9,770,000 |
| Biotica | 1996 | Cambridge | $13,553,000 |
| CDD (now biofocus) | 1997 | Cambridge | $4,420,000 |
| KuDOS Pharmaceuticals | 1997 | Cambridge | $40,760,000 |
| Cambridge Bioclinical | 1997 | Cambridge | $300,000 |
| Sense Proteomics (part of Procognia) | 1998 | Cambridge | $4,186 |
| Solexa | 1998 | Cambridge | $113,380,000 |
| Cambridge Cognition | 1999 | Cambridge/CeNes | $350,000 |
| De Novo Pharmaceuticals | 1999 | Cambridge | $4,500,000 |
| ImmunoBiology | 1999 | Cambridge | $9,070,000 |
| Spirogen | 2000 | London | $16,436,000 |
| Procognia | 2000 | Israel | $22,070,000 |
| SmartBead Technologies | 2001 | Cambridge | $531,000 |
| CellCentric | 2001 | Cambridge | $1,860,000 |
| Akubio | 2001 | Cambridge | $12,450,000 |
| Chroma Therapeutics | 2001 | Cambridge | $91,340,000 |
| Cambridge Biotechnology | 2001 | Cambridge | $553,000 |
| Vivamer | 2002 | Cambridge | $170,000 |
| Purely Proteins | 2002 | Cambridge | $3,760,000 |
| Smart Holograms | 2002 | Cambridge | $11,960,000 |
| Lumora | 2003 | Cambridge | $2,280,000 |
| Sentinel Oncology | 2005 | Cambridge | $190,000 |
| Phico Therapeutics | — | Cambridge | $810,000 |

SOURCE:  VentureXpert database; Thomson SDC Platinum Version.4.0.3.1.
NOTE:  This list includes only the companies that had venture capital information reported through VentureXpert (Thomson SDC Platinum Version 4.0.3.1).

The University of Cambridge's decline in spin-out formation since 2000 is in contrast to the spin-out activity at MIT, Oxford, and Yale. MIT, while not forming as many spinouts as it did during the 1990s, keeps spinning out two to three biotechnology companies a year. More interesting, Yale increased the number of spin-outs per year from three biotechnology companies in 1993 to seven in 2004, while in the same time period

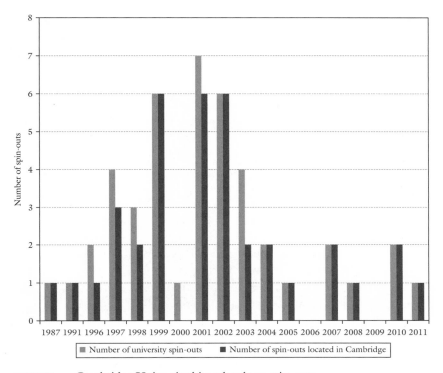

FIGURE 5.1  Cambridge University biotechnology spin-outs
SOURCE:  Cambridge Enterprise (2009, 2013).
NOTE:  Eighty-four percent of the biotechnology university spin-out companies chose to stay in the region.

Cambridge spun out one or two companies per year. Nor was the decline in biotechnology spin-outs in Cambridge due to national or international factors. Unlike the decline in biotechnology spin-outs at the University of Cambridge, nationally we find a growth of UK biotechnology industry from 94 in 1997 to 125 companies in 2003.[13] In particular, Oxford University spin-outs grew from one in 1993 to four in 2004. Furthermore, the global biotechnology industry grew from 4,284 companies in 2001 to 4,471 companies in 2003. Examining this trend continuously to 2011, we find that Cambridge continues to stagnate. Thus, although the numbers are small, it seems that the decline in the number of the University of Cambridge spin-outs was not part of the general trend in the industry but was the result of reasons endogenous to the university, such as the constant reshuffling of technology-transfer organizations. From interviews with local industry members we find that industry viewed the process of change in Cambridge as unprofessional, secluding, and confusing. The change created a situation in which executives in the region made deliberate choices not to work with the university.[14]

In summary, this section finds that while IPR policy went through vast changes to improve the university's ability to commercialize technology, the results were not constructive. After many struggles in the faculty senate, changes in policy gave more control to the central university administration. The changes in policy were concentrated on IPR ownership, while other technology-transfer policies—such as company location within laboratories, impacts on tenure and promotion, and the ability to hold dual positions—were not centralized, and decisions on those issues are made by individual departments. For example, there are vast differences among departments. In some departments, spin-out companies can be totally isolated from any departmental activity, whereas in others the company laboratory will be located on the same floor with free access to students. Faculty, although not encouraged to do so, can hold an official position both in the university and in a company.[15]

### Organization

The process of commercializing academic ideas in Cambridge reflects the collegiate system of departments and colleges at the university. It is conducted on many levels and in many organizations, which makes it, using Hatakenaka's (2002) word, "fuzzy." The university created some of the units dealing with technology transfer; some were created by colleges, and yet others were created by departments within the university. Furthermore, since the early 1990s, constant changes in the organizational structure, titles, and management have created an uncertain business environment. Many of the organizations have no knowledge of the others, and some provide duplicate services, which makes access to, and understanding of, the university technology-transfer policy difficult for academics, students, and industry representatives.

Technology Transfer Managed by the University's
Central Administration
Following the Mott Report of 1967 and the creation of the Cambridge Science Park by Trinity College in 1970, the Department of Engineering created its own technology-transfer unit. In 1983, this unit became

the Wolfson Industrial Liaison Office (WILO), responsible for the commercialization of university research as a whole (Minshall, Druilhe, and Probert 2004). WILO was a two-person organization: one person dealt with physical engineering and the other with biomedicine:

> We were facilitating. Because we didn't have many resources, we did what we did and no more and we were very keen on setting up collaborations, leverage, getting external people to want to do a deal and share the revenue with us, share the upside with us. That was great. Some people would say, "You're giving 50 percent away," and I would say, "It's at no cost to us, we get 50 percent of the benefit and no downside." When you're a small office you want to use as much of that leverage as you possibly can. So the whole approach was very collaborative, facilitative, catalytic, very much the art of the possible, or trying to make things happen. (Former member of WILO)

As a small office working with the entire university, WILO had advantages and disadvantages. On one level, all players knew the individuals at WILO and were discussing issues on a "director's" level. The directors understood they had limited time and availability to assist all firms. Moreover, faculty still owned the rights for their inventions. Many chose to go through WILO, but others didn't. Also, as mentioned by one of the directors: "We could not follow all patents or licenses. Hence, we made choices. We accepted self-reporting of royalties by firms. We did not have the funding to verify their reports." This method allowed faculty who chose to commercialize their technology to do so. It had less bureaucracy involvement and gave a free hand to individual entrepreneurs. However, WILO was not able to work with everyone and did not have the resources of a legal team and specialists. Hence, in cases where faculty needed help, or did not have the will or ability to promote an important technology, inventions were not developed.

Starting in 1999, the University of Cambridge began restructuring its technology-transfer units.[16] The reason for the change, which affirms existing organizational change literature, is government pressure and resources. Government pressure can be seen through the Dearing and the Sainsbury Reports, which emphasized the need of universities to contribute to the economy through technology commercialization. These reports led the UK government to provide additional funding for universities to promote university-industry relationships and technology commercialization.

Many of the organizational changes and new units handling technology commercialization at Cambridge were the result of a direct response to government funding. Hence, making WILO into a central university technology-transfer office and later combining all the units into the Research Services Division was a response to the Higher Education Reach Out to Business and the Community Funding Program of 1998. The University Challenge Fund (UCF) was created by the winner of the Higher Education Funding Council for England's (HEFCE) University Challenge Fund competition of 1998, and in 1999 the Cambridge Entrepreneurial Center was created as a response to the 1999 Science Enterprise Challenge. The center provided educational activities on the practice of entrepreneurship.

The Cambridge UCF, created in 1998, was a joint fund for Cambridge University and the Babraham Institute, with £3 million ($4.3 million) from the UK government. The UCF invested in spin-outs, to a maximum of £250,000. This amount of funding is not enough to spin out a company. However, it might be enough to assist companies in getting to a stage in their technology development at which they can recruit outside funding. By the end of 2004, UCF had funded fifteen companies, 50 percent of which were biotechnology firms. However, the fund's resources diminished over time, and the University of Cambridge refused to invest in the fund, leaving limited in-house option for university spin-outs. Hence, in 2008, as part of the eight hundredth anniversary campaign at Cambridge, the university established the University Discovery Fund, which funds spin-outs from the university for "proof of concept, pre-license, pre-seed and seed investments, enabling young companies based on Cambridge research to succeed" (Cambridge Enterprise 2009).

In March 2000 WILO merged with the Research Grants and Contracts Office to form the Research Services Division (RSD), creating one organization to deal with sponsored research from the research councils, industry, and charities. This new organization was headed by David Secher, formally an academic research scientist at the MRC Laboratory of Molecular Biology in Cambridge. Until 2002, several organizations operated under the umbrella of the Research Services Division. This includes WILO, which became the technology-transfer office (TTO), the UCF, the Cambridge Entrepreneurial Centre, and the Corporate Liaison Office, which supports partnerships with industry.

In 2002, to create a single organization dealing with technology transfer and because of government pressure in the form of the Higher Education Innovation Fund, the University of Cambridge created a new unit called Cambridge Enterprise. Peter Hiscocks, who previously managed the Cambridge Entrepreneurial Centre, headed the new organization.

Cambridge Enterprise was created on top of the Research Services Division, by removing and combining the technology transfer and the UCF from RSD.

In 2003, this change was reversed, and Cambridge Enterprise went back under the umbrella of RSD. At the same time, the Cambridge Entrepreneurial Centre was divided into two: one part joined the technology-transfer side of the university activities, and the teaching part became the Centre for Entrepreneurship Learning, now a part of the Judge Business School, headed by Shai Vyakarnam. The center's mission is to spread the "spirit of enterprise" by providing educational activities to inspire and build skills in the practice of entrepreneurship:

> Some of the most important are our teaching program and the business plan competition, which makes people and 50 students aware of what we do. Last year we taught 1,740 separate students one module or another. . . . If you take into account the number of students who come in a year, we are getting to about 20 percent of the students coming to a course. (Cambridge administrator A)

In August 2005, yet another round of restructuring was done, and Cambridge Enterprise became a single organization again, headed by a new director, Anne Dobrée. Dobrée left Cambridge Enterprise in February 2006, and in August 2006 a new director, Teri F. Willey, was appointed. She was replaced by Dr. Tony Rave in December 2011. In December 2006, Cambridge Enterprise became a limited company, a wholly owned subsidiary of the University of Cambridge. The organization includes the technology-transfer functions and the Cambridge UCF.

The mission statement of the organization focuses on the commercialization of technology. Unlike Yale, Cambridge's mission statement does not discuss local economic development or even benefits to society in general:

> Cambridge Enterprise's goal is to take the most promising ideas forward through intellectual property rights (IPR) licensing, new venture creation and consultancy. Throughout our work we aim to:
>
> - build strong relationships with University academics to encourage disclosure and cooperative management of the most promising innovations
> - make significant, measurable progress toward financial sustainability to drive long-term benefits to academics, Departments and University
> - be an attractive partner for industry and investors to take University ideas forward through commercial channels (Cambridge Enterprise 2013b)

Many of these changes in directors, splitting and consolidating units, cannot be explained by mere government pressure. Information regarding the reasons for the frequent changes in personnel cannot be officially verified. That said, interviewees implied there was a political power struggle within central administration and between central administration and the colleges as to the "right person" to direct technology commercialization in Cambridge.

Moreover, according to organizational change literature, the change itself can provide an explanation as to structural and policy changes within the technology-transfer office. Organizational change affects employees. Some find the change appealing and choose to stay, and others choose to leave. Especially in the case of Cambridge, the responsibility change for employees was vast. The volume of patents, licenses, and other agreements went up sharply as the university received intellectual property rights. Pressure to perform went up as well. Hence, employees felt pressure both from the university but now also from corporations and faculty who did not previously need the technology-transfer office's services. Moreover, until 2006, when Cambridge Enterprise became a private organization separated from the university, it was not able to pay competitive salaries.

In 2006, Cambridge Enterprise had eighteen employees, of which twelve had PhD's and fifteen had industry experience. By 2013 the organization had grown to include forty-eight employees (Cambridge Enterprise 2013a). The office includes the former corporate liaison office and contract negotiation group.[17] The changes in the University of Cambridge's technology-transfer organization are illustrated in Figure 5.2.

The constant structural, managerial, and name changes of the technology-transfer office at the university created the current situation, in which both faculty and industry are not sure who they are supposed to contact with regard to an invention or license. During the field research interviews, companies and entrepreneurs, when asked which organization they will turn to, gave a specific name, such as Walter Herriot (from St. John Innovation Centre, a college organization), David Secher (director of Research Services Division), and Richard Jennings (currently with Cambridge Enterprise and former director of WILO), rather than organizations. When specifically asked about organizations, interviewees' typical answer was "that organization . . . Cambridge something on Mill Road":

> As with any function, technology transfer in Cambridge comes down to individuals. The process helps but not much. I have good connections at the

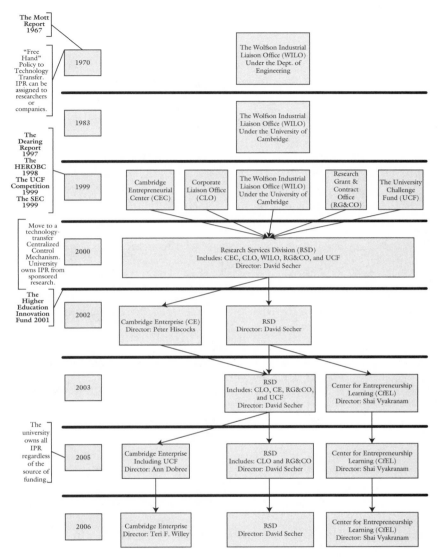

FIGURE 5.2 University of Cambridge technology-transfer organization,
1970–2006

NOTE: The constant organizational changes negatively impact the university's ability to commercialize technology.

university: David Secher, Geraldine Rogers, Bill Mathews, Richard Jennings. (Venture capitalist C)

A lot of the Cambridge success, I think, is due to these grand visionary individuals who have these long-term ambitions but no big corporate thing—you need to tell us who these folks are and how they did it! It's a collection of very high-quality individual initiatives. (Cambridge administrator C)

As these quotes show, the multitude of technology-transfer organizations at Cambridge created more reliance on key individuals who represent the university. Firms refer to the problem as "marketing issues," but the fact remains that many of the companies are still not aware that Cambridge Enterprise is the university's official technology-transfer office.

Moreover, faculty and firms had concerns about the level of professionalism with which technology transfer in Cambridge is handled:

[They move from] one extreme to the other. [First they had] one person overworked. Now [there are] more people and it is less facilitating. [They] were not that helpful. (Faculty member and company founder A)

The TTO gives the impression that it is the British civil service. [You] better not make the mistake and get the technology out [from the university]. (Biotechnology executive A)

People at the company felt the TTO in Cambridge was the most difficult in the world. Inexperienced members of the staff took large decisions without skills to work with a company with a different culture. (Faculty member and company founder B)

These statements show how critical and how doubtful both faculty and industry are of Cambridge Enterprise.

Technology Transfer Managed by Colleges and Departments

Technology-transfer organizations can also be found at Cambridge in some of the colleges. Cambridge's colleges are independent institutions, recruiting their own faculty and managing their own finances. In addition, several of the colleges have developed their own technology-transfer initiatives and organizations, thus contributing to the confusion over technology-transfer policies at Cambridge.

As mentioned earlier, Trinity College built the Cambridge Science Park in 1970. The initiator was John Brownfield, then the bursar of Trinity. Later, Brownfield explained that the Science Park initiative came about only because Trinity owned the land and was not using it. Since Trinity is very rich and did not need an immediate return on its investment, the college was able to go into a long-term investment, such as the Science Park, because, "well, if it makes money in 50 years, that's fine by us" (the words of John Brownfield, as recalled by one of the interviewees).

The park is located 2.7 miles northeast of Cambridge University, and it hosts seventy-one companies. The majority of the companies are involved in scientific research and development. There are also a few companies that directly support the research and development occupiers, and these include patent agents and venture capital funds. The park does not have an incubator space and caters to mature companies. Thus, it is considered more expensive than other parks, so most start-up companies cannot afford the location.

In 1983, St. John's College built the St. John Innovation Centre, which opened in 1987. In contrast to Trinity, the college was interested in a commercial return. It recruited Walter Herriot to manage the center, operating on three levels. The first are the tenants, about fifty mainly knowledge-based companies, with only a few university spin-outs. This is an office-based environment, not well suited for biotechnology, although some biotech companies have started there. The next level is the innovation center, with 192 members. On this level, the center is a virtual office for people who work from home, providing office space for meetings as well as administrative and logistics assistance. It has created a network of small businesses, in which the companies are smaller than in other organizations but have a newsletter and regular meetings. The last level works with publicly funded contracts, such as Gateway to Innovate, which encourages local companies to innovate. The center executes the government programs funded by the East of England Development Agency. The services are then offered to all companies in the region that fit the criteria of the specific program.

Other colleges soon tried to emulate the model set up by Trinity and St. John's. In 1995 and 1996, Pembroke College created the Pembroke College Corporate Partnership. The partnership is a membership program between corporations and the college. It is tailored to the requirements of the specific corporation, such as pursuing new relationships with a specific lab or faculty member or setting up and collecting information on research conducted at the university. To participate in the program, the college charges the corporation a fee according to its size and the sector of the company:

> Depending on the needs of the corporate partner, many different forms of interaction are possible, from the hosting of seminars, providing consultancy support, and establishing full-scale research programs in collaboration with the relevant university departments. (Pembroke college representative)

In December 2004, eleven companies were members of the program, three of which were also members of the Corporate Liaison Program at RSD.

Some Cambridge University departments have also created their own technology-transfer program. In 2000, the Department of Chemistry created the Corporate Associates Scheme. The scheme promotes a formal relationship between the Department of Chemistry and industry. Today, there are thirty members in the program. Like the Pembroke Corporate Partnership, the department charges its members according to the size of the company. Benefits of the scheme include information on research in the department, sources of potential collaboration, research lectures, teaching courses, and so on. The scheme also provides access to on-site recruitment of students, and to the department library and equipment (Department of Chemistry, University of Cambridge 2006).

An examination of the technology-transfer organization at the University of Cambridge uncovers a wide network of diverse offices dealing with the technology-transfer process. Central administration at Cambridge has one office that deals with technology transfer. Between the years 1999 and 2006, the office went through continuous changes. Changes were not done with a structured plan but more as a response to various pockets of funding, and thus they created confusion for both faculty and industry. This perplexity contributes to a business uncertainty that discourages industry from working with the university. To add to this confusion, the change was not inclusive. It was restricted to the central administration's technology-transfer units.

## LOCAL INDUSTRY'S VIEW ON THE TECHNOLOGY COMMERCIALIZATION CHANGES

The University of Cambridge, with about forty departments linked to biotechnology, provides a resource for the local biotechnology industry. This is strengthened by the notion that half of the university's government funding is dedicated to clinical medicine and biosciences (Minshall, Druilhe, and Probert 2004). Companies see the university as a source of technology, employees, and brandings. Interestingly, and as depicted in Figure 5.3 and the following quote, even after the vast changes in technology-transfer policy and organization, biotechnology companies view the University of Cambridge as important to the development of the biotechnology cluster:

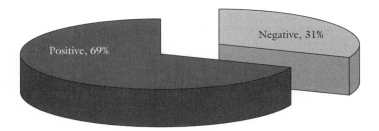

FIGURE 5.3  Biotechnology companies' view on the importance of the University of Cambridge to the development of the cluster
NOTE:  Interviewees' responses to the question regarding the university impact on the cluster.

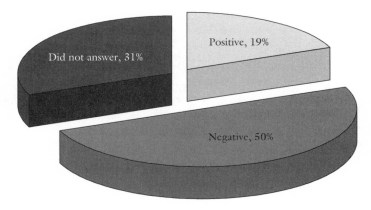

FIGURE 5.4  The biotechnology industry's view on technology-transfer changes at Cambridge
NOTE:  Industry does not appreciate the university's particular changes to its technology-transfer policy and organization.

> It is critical to the cluster. Provided the knowledge base, scientific talent. It was the catalyst for the cluster to happen. (Biotechnology executive G)

That said, industry does not appreciate the university's particular changes to its technology-transfer policy and organization. Indeed, companies found it very difficult to understand and navigate the technology-transfer organization in the Cambridge system. As represented in the following quotes and Figure 5.4:

> I think the TTO in Cambridge is behind. They don't do a good job publicizing themselves. The [licensing] process makes it difficult to raise money. (Biotechnology executive D)

> [We] never licensed from the university. The university is a good source of intellectual [property] but should not try to spin out. I don't know who to talk with about licensing. (Biotechnology executive A)

Even in universities such as MIT and Stanford, one always hears complaints about working with the technology-transfer offices and the difficulty in licensing technology. However, the University of Cambridge's problem goes beyond common complaints. In 2004, Cambridge had a reputation for having an unprofessional technology-transfer office that employed people who were not able to evaluate the technology in question and were not experienced enough in the business environment to understand what exactly a good invention is:

> The companies are formed in spite of them [Cambridge Enterprise]. Not on a business scale. You just need to look at the numbers. Look how much money gets funneled in Cambridge University. Only a handful. We did a few ourselves. IP to IPO in Oxford 30 million in one department and formed more companies in the last two to three years than Cambridge has done in a decade. This is an example of how it can be done differently. Cambridge University could actually leverage its own name, but it does not want to. There is no research consortium to learn about what is happening in the university. (Local venture capitalist and entrepreneur B)

> The process depends on the individual you work with at Cambridge Enterprise. Most of them had no experience in commercializing business. (Biotechnology executive E)

> I don't think Cambridge Enterprise has people with the proper experience. You need to develop networks. [Cambridge Enterprise] has high turnover of people, they can't be effective. (Biotechnology executive F)

Thus, companies have specific complaints regarding Cambridge Enterprise, its employees, and its ability to spin out companies.

Examination of the life sciences group at Cambridge in 2005 showed that employees were qualified academically but did not possess the business expertise required to evaluate new technologies. In 2005, of the seven employees in the life sciences group at Cambridge Enterprise, four had PhD's, of which two were in organic chemistry. Evaluation of the group's business background found experience primarily in management of research in an academic institute; except for the general managers, none of the employees in the life sciences group had held a management position in a pharmaceutical or biotechnology company.

The business community's criticism of Cambridge as having an inexperienced technology-transfer office due to its lack of business experience is corroborated through the work of Lockett and Wright (2005), which shows that the lack of business experience in the technology-transfer office has a direct impact on the capability of the office to spin out companies. Furthermore, other studies suggest that the lack of expertise in business skills, particularly in marketing, has a direct, negative effect on technology transfer (Clarysse et al. 2005; Siegel, Waldman, and Link 2003a). Accordingly, the fact that Cambridge Enterprise has very limited in-house expertise in evaluating the commercialization potential of new technology in biotech reduces the capability of the office to spin out companies.

## SUMMARY

In summary, the biotechnology industry finds the technology-transfer policy and structure at the University of Cambridge confusing. The multitude of organizations, the constant change of the organizations and directors, and changes in policy have created a situation in which, on many occasions, companies are not sure whom they should discuss technology transfer with at Cambridge. Furthermore, while attempting to work with the university, firms are relying on specific individuals to connect them to the correct place at the university. It is not surprising to find this in a cluster based on social networks with serial entrepreneurs. However, the fact that industry has such low esteem toward the university technology-transfer office is disturbing.

In the mid-1980s, entrepreneurs referred to the "Cambridge Phenomenon," a vibrant high-tech cluster around Cambridge. In the late 1990s, however, Cambridge initiated a series of changes in its technology-transfer structure and policy, and a biotechnology cluster in the area, one of the largest in the world, already existed. Although the university's goal of improving its technology-transfer model was intended to enhance its capabilities in this area, the changes instead unbalanced the delicate equilibrium that existed in the region.

Cambridge expected a positive outcome from its attempt to improve its commercialization. By rebuilding one centralizing technology-transfer office and investing more resources and accessibility to promote technology transfer, the university hoped to see an increase in technology commercialization. However, since the biotechnology cluster and social networks existed prior to the changes at the university, they were receptive to these changes, and hence relationships between local firms and the university were damaged.

In short, the changes in technology-transfer policy and organization had a negative result on the university's ability to commercialize technology. In particular, the university's choice to make partial internal university changes—changing the central organization of technology transfer while keeping all the other technology-transfer offices on the college and departmental levels; individualistic changes, without any coordination with other regional players; and constant changes, changes in organization structure, name, and directors over a long period of time—resulted in adverse results for Cambridge. While the university implemented many changes that according to existing studies should lead to improvements in technology transfer and commercialization, the implementation of those changes has been long and chaotic.

The following chapter compares the two case studies and discuss the implications for university technology transfer in particular and the role of the university in economic development in general.

## CHAPTER SIX

# *Apples to Apples*

The cases of Yale and Cambridge universities strengthen the argument that universities' ability to have a positive impact on their local economy varies. Yale's changes to technology commercialization led to the creation of a local biotechnology cluster. At Cambridge, which was the leader in the development of the local biotechnology cluster, technology commercialization changes damaged its ability to spin out companies and to collaborate with industry.

While both institutions strived for success, they achieved varied results. Those differences stem from the following:

- Different institutional and regional histories and environment
- Different technology commercialization culture, policies, and organization
- Different approaches to the process of change—specifically, vast differences in the velocity of change, the number of changes within the universities, and the universities' ability to collaborate with regional players

This chapter outlines the differences and commonalities of the two cases. We start with the external factors, followed by the internal university factors: policy, organization, and culture. The third section of this chapter reviews additional factors regarding organizational changes, which were crucial in these cases. Last, we examine Cambridge and Yale versus other models of technology commercialization.

## EXTERNAL UNIVERSITY FACTORS

As this book has shown, it is critical to understand how important the external university factors are, such as the history of the region and the system of innovation in which it operates, to the ability of universities to commercialize technology. It is clear that universities' ability to innovate is influenced by external factors in general, but specifically for technology commercialization, many internal factors, such as commercialization policy, organization, and culture, are significantly influenced by the environment and locale in which the university operates. The two cannot be separated.

Our two case studies indicate that legislation on the national or regional level has a direct impact on the ability of universities to commercialize technology. For example, in the United States, the Morrill Act of 1862, known as the Land Grant Act, and the Bayh-Dole Act of the 1980s have had a huge impact on the creation of technology-transfer offices (TTOs) as well as the commercialization pipeline of US universities. The Morrill Act had a direct impact on the commitment of universities to their regional economies. Being a land-grant institution meant a close and continuing relationship between the university and the local region. This commitment by universities resulted in an impact on local policy, job training provided by the university, and industry development based on university research. Data show that the number of TTOs in universities increased gradually following the Bayh-Dole Act, and today all US universities have technology-transfer offices (Sampat 2006; Feldman and Breznitz 2009). Moreover, the number of US universities' patents, licenses, and spin-outs has grown tremendously since the 1980s act.

Funding is another important factor that affects the ability of universities to commercialize technology. In the United Kingdom, it was only after the formation of the National Research Development Corporation in 1949 and the 1965 Science and Technology Act that UK universities started to receive government funding. Later, the creation of the research and funding councils, which allocate funding for research in general but also for specific research projects, affected the ability of British universities to conduct research. In the United States the creation of the National Science Foundation, the National Aeronautics and Space Administration, the National Institutes of Health, and other federal research institutes, such as the Defense Advance Research Projects Agency and the Department of Defense, proved vital for the progress of the country as a whole and US research universities in particular. In both countries, the national governments are the largest funders of university research.

Historical events such as World Wars I and II affected both countries and the way they viewed universities and their economic contribution. Moreover, national-level reports such as the Robbins Report in the United Kingdom and the Vannevar Bush Report in the United States had consequences for the abilities of universities to commercialize technology. Local events proved important as well. In Cambridge the Holford Committee and the Mott Report enforced the preference for research over manufacturing in the region. At Yale the death of a student on campus prompted the university to engage with the local region and promote economic development initiatives. These important events shaped the environments in which each university operates.

Regional culture and town-gown relationships matter. Cambridge's status as a top university was never connected to or thought to be affected by local policies and events. But for the purpose of understanding its technology commercialization capability, it is vital to remember that the town of Cambridge is small, and for many years it wished to stay that way. Hence, the county had strict regulations regarding manufacturing and even large research facilities. And until the 1990s, Yale, as a private elite university, viewed itself as separate from the local region and, as such, did not see the need to pour resources into the town of New Haven.

Last, the ability of a university to commercialize its technology lies in the local innovation system. Cambridge had a strong regional network of innovation and innovative partners. The region is home to several national laboratories, research councils, European institutes, and many firms, of which the majority spun out of the university. The social network in the region is very strong as well. While in other places such networks provide the basis for future innovation, for Cambridge, the existence of this tight network, which allowed the biotechnology cluster to flourish, worked against the university when it decided to change its technology commercialization framework. During the change, the university did not communicate with or consult local firms, including spin-outs. This lack of communication drew animosity from the community. Every firm or entrepreneur who had had a bad experience with the university shared it with other members of the social network. Many of the individuals I interviewed told of other colleagues who had had a bad experience with the university's technology-transfer process.

At Yale, the local innovation system was limited. Unlike Cambridge, with its varied networks of corporations and research institutes, New

Haven had a local innovation network made up mainly of large multinational firms that were not collaborating with one another or with the local universities. Hence, Yale had to be proactive in its attempt to impact the local economy. The university worked to strengthen connections to the city and state, as well as to local venture capital firms, and it even played an important role in bringing talent to the region, directly recruiting management for firms.

## INTERNAL UNIVERSITY FACTORS

A university's technology-transfer policy, organization, and culture are the most important factors affecting technology commercialization. This section compares the findings of the two case studies regarding these three factors.

### Technology-Transfer Policy

Universities implement their technology-transfer policies in different ways. The choices made by universities affect the way these institutions collaborate with industry, and hence lead to different results in terms of economic growth in the region, number of spin-out companies, and university-industry collaborations. The studies of Yale and Cambridge confirm the critical importance of issues like royalty share, spin-out policy, and faculty rights (Di Gregorio and Shane 2003; O'Shea et al. 2005; Roberts 1991; Shane 2004). Moreover, they highlight the importance of implementing technology-transfer policy changes over a short period of time and maintaining the pace toward the final outcome.

At Yale, policy changes began in 1993 when its technology-transfer office started to actively seek new inventions. The office made a procedural change to expedite the examination process and to focus commercialization efforts on the inventions viewed as those with the most potential for commercialization. At Yale and in New Haven the policy changes had a positive impact for both faculty and other regional players, who found the university to be a desirable partner in technology commercialization.

At Cambridge, policy changes were focused on intellectual property rights and centralization. From 1998 to 2005 the university continuously debated and made changes to its policies on intellectual property ownership. Cambridge moved ownership from inventors to the university. Moreover, many saw its new commercialization process as overly bureaucratic. The policy changes therefore had a negative impact on both university faculty and other regional players. Many firms that used to collaborate with university faculty and fund research at the university found that the changes limited their ability to work with researchers and to access research results. Faculty, especially those who had already been commercializing technology, did not see the need to go through a centralized office, and they were not happy to lose ownership of and control over their own inventions.

### The Organization of the University Technology-Transfer Office

The analysis of Yale and Cambridge finds support for several factors believed to affect a university's technology-transfer abilities. One, the organization of the technology-transfer office—that is, the number of TTO employees, their background, the use of outside lawyers, and so on—affects the ability of a university to patent and license its technology (Link and Scott 2005; O'Shea et al. 2005). Second, past successes encourage technology transfer and commercialization (O'Shea et al. 2005). However, at Yale and Cambridge it was mostly the process by which the change was conducted that had the biggest impact on the commercialization ability of their TTOs. Velocity and intensity of change, as well as inclusion in the process of change, were found to be critical components in the success of the technology commercialization effort at both universities.

Cambridge originally had a small technology-transfer office with two employees. Given its size and hence its ability to provide service to all faculty, students, and local firms, the office had a relaxed attitude toward technology commercialization and focused more on bringing technology to the market and less on the bureaucratic process by which commercialization was done. The technology commercialization change that started in the late 1990s created the current technology-transfer office, a separate entity called Cambridge Enterprise, a large office with specific regulations and procedures to commercialize technologies. Moreover, while Cambridge Enterprise has the academic expertise to evaluate many of the technologies, it lacks in commercialization and industrial experience. This lack of business experience had a negative impact on the development of the local biotechnology cluster and the commercialization efforts of the university. Although over time the university spun out thirty-seven biotechnology companies, the TTO had not succeeded in bringing a "big" success in producing royalty licenses or patents, or a large biotechnology company in the region. Importantly, not only were the organizational changes in Cambridge part of a long process; the TTO's composition, employees, and directors constantly changed throughout that process. Any organization going through so many changes over such a long period of time becomes significantly less effective.

Yale represents the exact opposite scenario. Yale's changes to its technology-transfer office were implemented in a short period of time, between 1993 and 1996, with the university undergoing an all-around change to promote technology commercialization, and continuously communicating those changes to the local industry. The university created a professional and engaged technology-transfer office, the Office of Cooperative Research (OCR). The office moved from being very bureaucratic to being business oriented, with its main goal the commercialization of technology. The results of its efforts can be seen in a vibrant biotechnology cluster, a growth in commercialization output, and a positive industry attitude toward Yale and the OCR in particular. Moreover, Yale has had several "blockbuster" inventions that have affected the way the university and the OCR view technology commercialization.

## University Technology Transfer and Commercialization Culture

University culture influences university-industry relationships, technology transfer, and commercialization capability (O'Shea et al. 2005; Saxenian 1994; Shane 2004). The stories of Cambridge and Yale support this notion. Importantly, those stories indicate a strong connection between the environment and history of a region—that is, external university factors—in which the university is situated to the existing commercialization culture. Hence, the way technology commercialization is viewed in each institution has roots in both historical events and the innovation system of the region.

As one of the leading universities in the world, Cambridge views itself as a national and international university—but it does not have any planning or outreach programs for the local region. Officially, its mission statement reflects its attitude toward local economic development. The university sees no specific obligation toward the town of Cambridge or toward Cambridgeshire County. Hence, its commercialization policy is general, it does not promote local spin-out formation in particular, and it does not specifically engage with the local industry. Interestingly, many of its spin-outs chose to stay in the region and have formed a social network of "Cambridge firms" (Garnsey and Hefferman 2005; Myint, Vyakarnam, and New 2005). The university historically had a supportive culture toward industry and applied research. In the words of the authors of *Cambridge Phenomenon*, "The University's consciously relaxed attitude and the simplicity of its industrial liaison arrangements have together helped nurture a 'culture' that encourages and is supportive of links with industry" (Segal Quince Wicksteed 1985, 69). This culture was the basis of the previous technology-transfer policy at the university. The change at Cambridge, which did not consider the regional connection as important, damaged these social networks and crippled the ability of the university to commercialize technology.

Yale University had to work very hard to change its previous culture, which was not supportive of technology transfer. This may be the most important aspect of the university change for Yale. Historically known as a university that did not care about its local region in general or technology transfer in particular, Yale realized that it needed to begin with cultural changes. The changes started with personnel changes at all levels of the university, from the president's office to the TTO. Yale hired individuals who were known for or made clear their commitment to local economic development and technology commercialization. Yale's new president, Richard Levin, was an economics professor at the university known for his favorable approach to economic development. The head of the new Office of New Haven and State Affairs, Bruce Alexander, had been involved in several urban developments around the United States. Moreover, the OCR approached individual faculty and researchers at the university to assure them that they had partners at the university. In addition, Yale worked to establish local firms, collaborate with existing local pharmaceuticals, bring in venture capital, and build companies. Its efforts paid off with the creation of a vibrant local biotechnology industry.

## CONTRIBUTION TO ORGANIZATIONAL THEORY

So what can we learn from Yale and Cambridge? Following the cases of Yale and Cambridge provided a rare opportunity to examine and compare cases of organizational change. In particular, the way in which universities choose to make their changes—the intensity of changes, the velocity of changes, and the inclusion of regional players—is crucial to technology-transfer efforts. Cases of technology commercialization need to draw from organizational studies when evaluating organizational changes.

### Intensity

Change intensity refers to the university's ability to make a comprehensive technology commercialization change within the university. It's not enough to change intellectual property rights (IPR) policy or the technology-transfer office itself. Any organization that works with industry and has an impact on technology commercialization must be included and become part of the new technology commercialization vision of the university. Top-down decisions are not enough. Grassroots movement among faculty, students, and staff has to be created.

Yale University made a high-intensity change in its internal culture. The university moved from a standoffish attitude to embracing commercialization as a solution for a university and regional crisis. The change was broad, including technology-transfer culture, organization, and policy. Furthermore, changes involved all the university departments, as well as administrative offices and officers. In particular, the university worked very hard to change attitudes and understanding of employees who had been working at the university for a long time. The commitment of the university to the change was also high, as demonstrated by the funding and personnel it allocated to this venture. The process involved everyone at the university, from the president to the most junior faculty.

Cambridge University's technology commercialization change start point was different. The university already had an internal culture of commercialization, based on its long-standing relationship with the region, which had a vibrant biotechnology industry. Still, the university technology-transfer process was diffused among departments, colleges, and central administration. Moreover, the changes to technology transfer were made as a reaction to governmental reports and pots of money. Hence, changes were made only at the central administration level, within the technology-transfer office and to the university IPR policy. The changes were not co-ordinated with or communicated to other university technology-transfer entities (at both departments and colleges) and to interested parties outside the university, such as venture capital firms, science parks, and local firms. What Cambridge lacked, therefore, was intensity of change, which ultimately reduced the effectiveness of its technology-transfer changes.

*Velocity*

The velocity of change is critical. How long does it take to make the necessary change? Are we talking about weeks, months, years? Is change done in waves? Do people have time to understand the change, or are we constantly changing, causing confusion?

Yale University's change took just three years. The start point of the change was very clear and was declared at both the university and regional levels by the university's president at the time. The university had clear goals to achieve through its overall activities, and they were stated in advance. To make the necessary transformations, the university invested its own funding and made elaborate changes to technology transfer at all levels of the university. The change was complete when these goals, especially local economic impact, were achieved.

At Cambridge, change happened in random and unplanned waves. The first wave of changes started in 1999 in response to streams of government funding that stirred up different issues regarding technology commercialization and entrepreneurship. Subsequent waves of changes were responses to internal university changes due to bureaucracy and politics, as well as the receipt of additional funds, such as the Higher Education Innovation Fund of 2001. The change in technology transfer stretched out over seven years. Therefore, this study reveals the importance of the velocity of change. One wave of change is more effective than several rounds of changes over long periods of time.

*Inclusion*

When Yale University decided to make economic development changes, its leaders were certain of their need to include other regional players—the State of Connecticut, the City of New Haven, regional firms, and entrepreneurs—to achieve their goals. It was the university's view that its isolated attitude toward the region and the city was partly to blame for its need to change. The university had missed the high-tech boom of the 1990s, its buildings (especially property it owned in the downtown area) were in bad need of renovation, campus security needed improvement, and there was a lack of a regional industry in life sciences. Change was imperative, and the first step toward that change was the creation of a new office, the Office of New Haven and State Affairs, to support collaboration among the university, New Haven, and the State of Connecticut.

Moreover, the changes at Yale were done in communication and collaboration with the State of Connecticut, local pharmaceutical companies, start-up companies, and venture capital firms. All players were aware of the changes and had the opportunity to collaborate with the university in its attempt to affect local economic development. When the majority of the changes were done, local pharmaceutical firms had created official collaboration channels with the university, venture capital firms were investing in spin-outs, and the State of Connecticut had its own venture capital arm that made a point of investing in local firms. Yale's policy of inclusion was one of the key factors in its success.

In contrast, Cambridge's technology-transfer changes were internal to the university and were done without including other regional players in the discussion. Through its waves of technology-transfer changes, the university did not consult or collaborate with its own spin-outs, local firms, and other research institutes in the region. The impact of such one-sided actions created animosity in local firms to the point that they refused to collaborate with or support research or university spin-outs.

This book emphasizes the importance of environmental factors on university technology-transfer success. Making sure that regional partners are included in a university's commercialization change is crucial. Universities do not operate in a vacuum. Their relationships, communication, and collaboration with government agencies, firms, and other regional players are vital for their ability to commercialize technology. Changes in policy and organization at a university need to take these players and their connections to the university under consideration.

## WHAT DO WE KNOW ABOUT OTHER TECHNOLOGY-TRANSFER OFFICES?

Increasingly, university and regional administrators are looking to create the perfect technology-transfer office. If possible, university administrators are looking for a specific formula; as Anne Miner and colleagues (2001) noted, we constantly look for the "secret sauce" to build the perfect technology-transfer office. However, as this book shows, there is no such thing. What we can do is identify and analyze important factors that need to be taken into account and best practices.

When we talk about a "model" technology-transfer office, we automatically think of MIT and Stanford, two of the top commercializing universities located in two of the most innovative regions. However, MIT and Stanford have what Roberts and Malone (1996) call the "low selectivity, low support" model. Offices like theirs commercialize the most technology but provide little support. This model fits regions like Boston and the Silicon Valley very well—regions with a rich base of venture capital and serial entrepreneurs who constantly search for new technology and can provide the support firms need. University technology-transfer offices in those regions do not need to hold the hands of faculty or students while they attempt to develop their firms, nor do they need to provide advice on how to create a company or network with other professionals. The regional innovation system is developed enough to provide market-based solutions for these issues.

Founded in 1861, MIT is a land-grant university. It was created with a mandate to link with industry. Hence, from inception this institute of technology placed importance on innovation, entrepreneurship, and local economic impact. The Technology Licensing Office (TLO) is considered one of the top performing, with an average of more than ten spin-out companies a year. The reason for this success, according to its director Lita Nelsen, is the region in which the university is situated:

> I attribute the difference [MIT's commercialization strategies to Yale] to the fact that there wasn't much infrastructure in the New Haven region, whereas in the Cambridge region this infrastructure has built up over time. So when people ask whether MIT has an incubator? Yes, it is the city of Cambridge[;] it is a geography experienced in high-tech entrepreneurship.
>
> The greater Boston area has a long tradition of starting up technologically ventures from their universities and also spinning out of corporations. It has trained lawyers, experienced accountants and real estate agents. We also have indigenous VC firms and management "know how" who demonstrated capability to build and launch startups from university labs. This has also generated a feedback loop within the regions, whereby clusters feedback themselves and a "success breeds success" culture has emerged. (Breznitz, O'Shea, and Allen 2008, 139)

The reality is, however, that most universities are not situated in such a vibrant and supportive environment. Yale University hopes that in the future it can operate in a model similar to that of MIT, with very little support, but the New Haven region does not currently have the resources that technology commercialization needs. There are not enough entrepreneurs, management, or venture capital firms. The university itself and the role and importance that technology commercialization have at the university matters. The fact that a university is a public or a private institution affects decision making within the university and its commitment to the local region.

Consider the Georgia Institute of Technology (Georgia Tech), for example. Even though Georgia Tech is not a land-grant university, it was established to make an economic development contribution to the local economy, which is still one of the main missions of the university. Hence, technology commercialization is an important aspect of the university. This is evident by the prominence technology commercialization receives in the amount of inside university funding, the connections to the State of Georgia, and the fact that technology commercialization directors report directly to the vice president of the university.

As a university that is committed to technology commercialization, Georgia Tech created the Enterprise Innovation Institute to collaborate with industry and commercialize technology (Youtie and Shapira 2008). This institute has been named as one of the top university incubators in the world (Steiner 2010). Moreover, in 2009, Georgia Tech was ranked one of the top ten patenting universities according to the Patent Scorecard, which evaluates the use of technology by industry.[1] Comparing the university with the other top ten universities on the 2009 scorecard, Breznitz and Ram (2013) found that for a relatively young technology-transfer office (1990) and relatively low R&D expenditures, Georgia Tech compared very well. This was due to the importance that technology commercialization is given at Georgia Tech.

Being a public university can create barriers for technology commercialization and local economic development. The level of resources available for universities to nurture entrepreneurship and commercialization (both are not an obvious role of universities) affects their performance. Public universities are confined by pay levels and hiring policies that many private universities are not. Some universities have solved these issues by creating a private entity that manages its technologies and liaison with industry. Oxford, located in one of the most research-intensive regions, provides ample funding and support for its entrepreneurs through its private technology-transfer office, ISIS. Created in 1987, ISIS is a private company wholly owned by the university (Lawton Smith and Ho 2006). As such, the office has eighty-three employees, with forty-two PhDs and fourteen MBAs among them. The result of the office professionalism is evident in its ability to commercialize technology. It has spun out eighty-five companies and since 1988 averages five spin-outs a year.

Factors pertaining to the quality of the technology-transfer office itself have been found important in both this and previous studies. The quality and amount of office personnel, especially their business experience, have been found to be a crucial factor in the success of a technology-transfer office. There must also be adequate funding available to hire people with PhD's who also have business experience. Universities need to compete with large corporations, which is very difficult. However, offices that have business experience have been able to evaluate the technology and estimate the amount of time and effort that need to be invested in the technology to get to the market, and they are also able to negotiate with firms, venture capitalists, and academic partners. Moreover, offices that have seen a technology move from invention to market, bringing in financial returns, are motivated and have better understanding of the commercialization process; as such, these offices are better able to bring more inventions to the market. The TTOs at Yale and MIT are good examples. Yale has several successful drugs, with the most famous one being Zerit, licensed to Pfizer, which brought billions back in revenues to the university. MIT saw many ideas become successful companies, such as Luminus Devices and Akamai. Stanford technology and graduates created corporate giants such as Google, HP, and Yahoo.

University policy regarding technology commercialization is also important. Policies addressing IPR ownership, royalties' distribution, support for junior faculty members, equity for patenting and licensing expenses, exclusive licensing agreements, the ability of faculty to take a leave of absence, and the ability of start-ups to use university resources all have a direct impact on the ability of a technology-transfer office to commercialize technology. Access to pre-seed-stage capital is equally important.

Existing literature and my own previous research on different technology-transfer offices, such as those at Georgia Tech, Emory University, MIT, Yale, and Hebrew University, have found that each university has a different formula of both internal and external factors that affect technology commercialization. The success of these offices essentially comes from how they work within the framework of their own history and environment. For example, a public university that cannot afford to hire professional technology-transfer officers cannot evaluate its technology and will most likely be unable to enter professional business negotiations with firms. Hence, it will be in its best interests, even if not always possible, to hire outside lawyers to conduct negotiations.

The comparison of Yale and Cambridge highlights the importance of analyzing technology-transfer offices with an understanding of their national and regional perspectives, their access to resources, and their general contribution to economic development. Technology-transfer offices play an important role in commercializing technology. However, we should not expect them to do the impossible. Considering their resources, their abilities, and regional support, we should provide them with appropriate tools to evaluate and commercialize technology.

## CHAPTER SEVEN

# Conclusion

The research presented in this book emphasizes the impact of location on universities' technology-transfer capability, and hence their economic development contribution. The region in which a university is situated, its history and environment, is critical in the way it influences internal university mechanisms for technology transfer. Moreover, much of the output generated from technology transfer depends on the resources allocated by the university, local, and national governments for research as well as the technology-transfer office and related activities.[1]

Importantly, the stories of Yale and Cambridge demonstrate that through the commercialization of research, universities have the potential to make a positive contribution to local economic development. However, not all will, and not only because of their own decision-making process. The factors that affect the local economic impact of universities are greater than what the university can control by itself. Cambridge and Yale, though very similar academically, are situated in different parts of the world, bearing different culture toward technology commercialization, having access to different resources, and having different organizational composition and capability to deal with technology transfer. This is why they had such a difference in technology output and different contributions to regional economic development.

Moreover, this book has shown that universities that are attempting to improve their technology-transfer capabilities need to consider three organizational change factors: velocity, intensity, and inclusion of other players in the change process. In other words, the way in which a university goes about improving its technology-transfer capability matters. It is possible to change university technology commercialization ability and have local economic impact. However, the change process has to be thought through, and local issues must be taken into consideration. To have a positive impact, a university needs to make a change that includes all sections of the university, including culture, policy, and organization. Second, by choosing a particular path of change, the university also changes its role and its ability to contribute to the region. Both Yale and Cambridge made technology commercialization changes, but only Yale succeeded in creating a biotechnology cluster and making a regional economic impact.

## THE FOUNTAIN OF KNOWLEDGE

The main focus of *The Fountain of Knowledge* has been to investigate the differences in universities' abilities to disseminate academic ideas to the private market. In particular, the book places emphasis on the impact of universities on local economic development by examining empirical questions such as, How do universities change their technology-transfer abilities? How do those changes affect universities and their relationships with local industry? Does industry share the view that universities' changes in technology-transfer policy, organization, and culture have improved their ability to collaborate with and contribute to local industry?

To answer these questions, Chapter 2 reviewed existing literature on technology transfer, finding that most experts view universities' technology-transfer efforts positively and unconditionally, without considering the possibility that some change in universities' technology-transfer processes may have neutral and even negative results. Moreover, the literature emphasizes specific factors that may contribute to increased university commercialization, including policy, such as royalty share and faculty guidelines, as well as organizational issues, such as business experience and education of employees of the technology-transfer office. However, the case studies in the literature provide little empirical analysis of any university over time or through changes to its technology-transfer processes.

Existing theories assume that universities, by their mere existence in the region, can, should, and will contribute to regional economic growth (Breznitz and Anderson 2006; Clark 1998; Etzkowitz 1995, 2002; Jaffe, Trajtenberg, and Henderson 1993; Keeble 2001; Lawton Smith 2006; Porter 1990; Saxenian 1994; Wiewel and Perry 2008). However, not all universities excel in technology commercialization and collaboration with industry.

Policy makers and university administrators should analyze the influence of historical, environmental, and intrauniversity factors that affect their ability to commercialize technology. As we have seen, their actions and changes in policy and organization do not always result in the intended consequences. Accordingly, while most studies view university contributions favorably, I argue that universities' impact can be negative on commercialization output and collaboration with industry, and that we need a nuanced understanding of their role in technology transfer rather than assuming that more (university involvement in commercialization) is always better.

Chapter 3 described the historical and national framework in which the studied universities operate. This chapter found that the universities had different relationships with industry in the two countries. In the United States, higher education institutions had strong industrial ties from inception, whereas in the United Kingdom such institutions started to develop ties with industry only in the past two decades. Importantly, regional economic contribution has been an inseparable part of American universities, a part that UK universities have just started to grapple with in the past decade.

Chapters 4 and 5 described technology transfer, organizational changes, and local economic impact at the two universities. The findings from the two cases show that universities can change their ability to disseminate academic ideas to the private market. However, not every intrauniversity change will have positive results. As we have seen from the case of the University of Cambridge, some changes can damage the ability of a university to commercialize technologies. Depending on the way "improvements" to technology transfer are done, organizational change can damage the commercialization capabilities of one of the best-performing universities.

The case studies of Yale and Cambridge have emphasized how important it is to realize that universities are heterogeneous, complex organizations. A close examination of each of these cases identified different approaches to technology transfer and commercialization. Each has advantages and disadvantages. There is no "secret sauce" or a silver-bullet model that one can apply. Each university needs a tailor-made approach to commercialization, one that relies on a regional history, culture, economic, and research capability. No two universities are exactly the same, serve the same population, or have the same history and culture. Hence, no two universities should have the same model of technology transfer.

Importantly, the mere existence of a technology-transfer office is not a guarantee of technology commercialization.

## THE FUTURE OF UNIVERSITY TECHNOLOGY COMMERCIALIZATION

Since the enactment of the Bayh-Dole Act in the United States in 1980, the debate over the merits and disadvantages of technology commercialization at universities has flourished. For many the Bayh-Dole represents what is known as the "third role" of universities, technology commercialization. In other countries around the world, similar laws and acts have been created to encourage technology commercialization, which has become a cornerstone for innovation and economic development. The third role of universities has led to the establishment of university technology-transfer offices and an increase in the number of universities' patents, licenses, and spin-out companies (Feldman and Breznitz 2009; Sampat 2006; Thursby and Thursby 2003). The addition of this role creates a responsibility for universities to make a return on investment to society by making sure that important discoveries get commercialized. As generators of innovation, universities are brought to the forefront of local economic development.

Criticism of university commercialization is not new. University-industry relationships in general and technology commercialization in particular created two sets of criticism. As early as 1977, higher education institutions were criticized in general for providing service to industry and for losing their identity as institutions of free spirit and open horizons (Scott 1977; Russell 1993). Later, there was criticism of the impact of commercialization on academic freedom, as well as criticism of the technology commercialization systems at universities (Krimsky 1987, 1988; Kenney and Patton 2009; Litan and Mitchell 2010; Litan, Mitchell, and Reedy 2007). The main complaint has been that universities are losing their identity. Krimsky (1988) claims that the connection between university and industry alters science per se by connecting academic research to corporation needs and financial gains.

While this claim is worrisome, no evidence has been found to support it. Studies such as the one conducted by Thursby and Thursby (2005) show no change in either research direction or financial return for research focus. Similarly, many of the faculty I interviewed at Cambridge and Yale rejected the notion that their focus on applied research is affected by sponsorship. Furthermore, research faculty in both universities stressed that funding *follows* initial research.

A second front of criticism attacks technology-transfer offices for becoming bureaucratic and for not providing "enough service" or "the correct service" for industry by authors who believe that commercialization of technology should be done by a professional service, or that ownership should be transferred to the inventor for commercialization purposes (Kenney and Patton 2009; Litan and Mitchell 2010; Litan, Mitchell, and Reedy 2007). Litan and Mitchell (2010, 53) describe university technology licensing offices as "bottlenecks of technology." Adding market freedom to the discussion, Litan and Mitchell argue that professors should be allowed to choose the agency with which they would like to commercialize their technology. They believe that university licensing offices should strive to improve service and commercialization output or maybe dismantle technology licensing offices altogether (Litan, Mitchell, and Reedy 2007).

Kenney and Patton (2009) support Litan and Mitchell regarding the influence of the Bayh-Dole Act on the commercialization of technology, claiming that the existing university technology commercialization model is not optimal. While the Bayh-Dole Act's purpose was to promote knowledge transfer and commercialization of technology from higher education institutions to industry, the actual result is a bureaucratic system that delays technology diffusion through ineffective incentives and revenue-maximization goals (Kenney and Patton 2009).

Although both sets of criticisms have some merit, many issues relating to the history and commercialization culture are not being considered. As described in this book, success in technology commercialization is based on many factors, and some have direct correlation with resources. These factors are due mainly to the knowledge and experience of the technology-transfer office personnel, as well as to the office's ability to hire outside lawyers and accountants. As in the case of Cambridge in the United Kingdom, many universities are public research institutions. Hence, technology-transfer employees are considered administrative support staff, and as such, they are compensated accordingly, meaning that the level of skill required from employees in those offices exceeds the pay offered. Moreover, registering and defending patents is expensive. The majority of universities do not possess the capability or resources to follow patent protection through. Importantly, as happened in Cambridge, some public universities addressed the resource issue by creating a professional private organization that provides technology-transfer services to the university but is self-supported and hence has full control over its resources. This is the case with the University of Wisconsin's Wisconsin Alumni Research Foundation, the Hebrew University in Jerusalem's Yissum, and Oxford University's ISIS Innovation, all of which are wholly owned technology-transfer companies.

When evaluating technology commercialization at universities, we must consider benefits stemming from those offices that are not as easily quantifiable as licenses, patents, and spin-outs. For example, knowledge transfer does not end with patenting and licensing. The addition of technology-transfer offices in many universities allows them to provide business assistance services that promote local economic growth and aid the universities' original role as promoters of new knowledge and social development. Pennsylvania State University offers consulting services and educational programs to entrepreneurs looking to start or grow small businesses. Virginia Institute of Technology's Technical Assistance Program, offered through its Continuing Education unit, connects businesses with expert faculty.

Further business assistance is delivered through incubator and science parks developed and located in close proximity to universities.[2] The University of Arizona Bioscience Park provides a separate facility designed specifically for companies working in biosciences, biotechnology, life sciences, and pharmaceuticals. MIT leased the land on which University Park at MIT was built. Today the park houses mainly biotechnology and pharmaceutical firms.

## REFLECTIONS ON POLICY

The cases of Cambridge and Yale present internal and external policy implications. First are the policy implications for universities. How should other universities frame their technology commercialization, innovation, and entrepreneurship policy to make local economic impact? Second are the national and regional governmental policies for funding universities. How should those policies be framed, and what is a reasonable expectation of universities?

### University Policies

Many universities around the world grapple with problems similar to the ones Cambridge and Yale faced. While it is evident that universities have the ability to disseminate academic ideas and commercialize technology, the particular process and its timing make a difference in the outcome of these changes. Many universities need to improve their technology-transfer organization and policy, but change cannot come simply from copying the Stanford or MIT model.

This study highlights the fact that each university is unique in its ability to commercialize technology. Any model that a university chooses to follow needs to fit that university's characteristics, regional environment, culture, history, and the resources available for commercialization. A liberal arts college is not likely to spin out biotechnology firms, nor are teaching institutes that do not focus on research likely to invent the next Internet.

Second, changes to the university technology-transfer policy and organization need to take place in collaboration with other regional players, including industry and regional agencies. Cooperation with other regional players places any policy or organizational change in a regional context. This context provides the university with guidance to what technology commercialization mechanisms will work best in the region to achieve the ultimate economic development impact. To properly evaluate university actions, we must consider industry perspectives in these efforts. Thus, we must also look at industry knowledge of and participation in universities' activities directed at technology transfer and commercialization. Universities can create programs to support the transfer of technologies to the private market; however, those programs will not generate the desired impact unless industry finds them useful and accessible.

Third, technology commercialization needs to become part of what universities define as research excellence and not solely as a means to make a profit. If technology commercialization is to be sustainable, it needs to become part of a faculty promotion and tenure process. Junior faculty in particular are wary of engaging in risky activity that may cost them their position.

*Government Policy*

One clear difference between Cambridge and Yale that became evident during this research project was government policy. These cases represent different approaches toward universities as an important economic and knowledge source. On the national and regional level, the United States always viewed universities as an integral part of its economy, as both a knowledge source and a labor source. The United Kingdom, in contrast, especially with its long history in which universities were created to educate the elite, took a long time to understand the full economic potential its universities present.

Public policy toward universities needs to be examined both on the funding side and on the evaluation side. Although studies show that research funding has a positive impact on university output in the form of publications, patents, and licenses (Lawton Smith and Bagchi-Sen 2012), not all universities can make an economic impact. The same amount of research funding at universities with different strengths and disciplines will not result in the same level of innovation. Hence, in considering both funding and evaluation of universities, one needs to consider the institution, its purpose, and the desired result. We cannot use the one-size-fits-all model, as we currently do with technology transfer. For example, the expectation that any higher education institute needs a technology-transfer office is unreasonable.

Moreover, policy makers need to realize that they cannot measure their return in currency alone. The experience of universities such as MIT and Stanford, which played a central role in the success of their respective regions, teaches us that universities' return from patenting and licensing does not even come close to covering the expenses of basic research. In addition, universities' contribution to local economic development extends beyond the technology-transfer office. As discussed in Chapter 6, universities are generally active in their communities. They provide policy and economic support that extend from other activities that are not related to technology transfer. Trying to impose market values on universities' regional impact is difficult to impossible.

## CONCLUDING REMARKS

The research on which this book is based began in 2003, with a sense that there was a need to emphasize the importance of universities' role in economic growth. The starting assumption was that when a university invests in technology transfer and builds a relationship with local industry, it can have only positive results. However, as I was conducting the research for this book, it became clear that the opposite is true: not all universities' investments in technology transfer yield positive results.

In addition, the findings of this book highlight the fact that universities do not operate in a vacuum. External university factors have a direct impact on the ability of universities to commercialize technology.

Moreover, any attempts to improve or change technology commercialization at universities need to take the university's environment and region into account.

Third, universities are intricate and varying institutions. Hence, internal factors such as institutional policy, culture, and organization influence their ability to disseminate academic ideas to the private market. Although we would like to think that all universities are the same, their differences are broader than the topics they teach or the faculty they employ. Universities have different histories, cultures, and structures that affect the way they interact with the region in which the university resides, the way they view technology commercialization, and local economic development.

Many other factors and players contribute to regional economic success; all the factors and players that contribute to regional success must work together to achieve positive results. Universities were created not to promote economic growth but rather to educate and conduct research. Though easy to measure, economic contributions should not ultimately be measured in terms of capital. The existence of higher education institutions should be valued as a whole, as the source for education, research, technology commercialization, policy, and economic development.

Although all of these additional activities are important, research and education must be protected as the main activities of universities. We must remember that many of the world's most famous inventions were created at universities working on basic research. Thus, universities should have the freedom to continue to explore, without being confined in their research efforts. If there is anything to learn from today's global competition, with its ideas such as the knowledge economy, competitiveness, and creative economies, it is that we must continue to invest in higher education. Groundbreaking research and well-educated students are cornerstones of every successful economy.

As a result of the contribution of university intellectual property during World Wars I and II and the Cold War, and now with growing globalization and global competition, universities increasingly find themselves viewed as the answer to economics concerns of policy makers. There is constant pressure on universities to work on applicable research and to focus their teaching and research efforts in ways that have a direct impact on their local regions.

The question remains, though: Are we expecting too much of our universities? Universities were created as havens of free thinking. Institutional achievements were based on free academic thinking. The entire notion of a university and tenured faculty was to allow researchers to work on the most cutting-edge work without any financial, political, or ideological pressure. Yet universities are in a bind: just as shareholders expect revenues, we expect universities to think outside the box and make a return for public investments in higher education. When we do so, we ignore not only universities' main roles of teaching and research but also their many contributions to society, most of which are not quantifiable.

Universities are *fountains of knowledge*. As such, they need to be encouraged to continue to do what they do best: teach and conduct research. Universities can continue to produce cutting-edge research and scholars while making a positive contribution to their local economies—but the economic contribution should not be their main mission.

# Organizational Charts Indicating Technology-Transfer Organizations

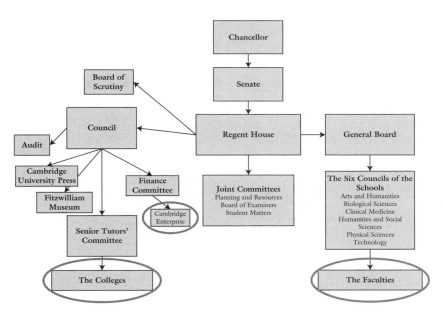

FIGURE AI.I  University of Cambridge organizational chart
SOURCE:  Lester (2007) and author.
NOTE:  The gray ovals mark the location of technology-transfer units.

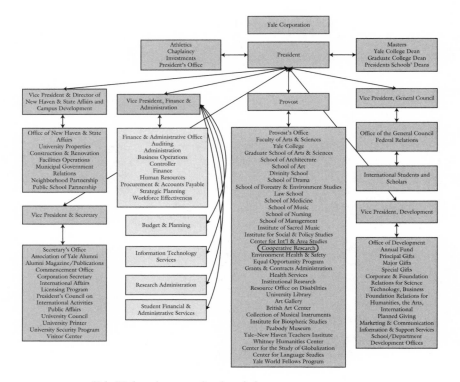

FIGURE A1.2 Yale University organizational chart

SOURCE: Office of the Secretary, Yale University.

NOTE: The gray oval marks the location of the technology-transfer unit.

# *Methodology*

*The Fountain of Knowledge* studied the impact of universities on economic development. The research project analyzed the economic impact of two universities through their technology commercialization activities. The research presented in this book is based on research conducted between 2003 and 2008, with data updates through 2013.

This study utilizes a case-study method, a useful way to conduct social research (Yin 2002). Unlike quantitative methods, in which large amounts of data are collected to describe a specific phenomenon, a case-study method conducts in-depth inquiry into a small number of cases to explain why a specific phenomenon has been found. In particular, Yin (2002) claims that case studies examine "real-life" events, and can include quantitative evidence, while relying on multiple sources of data.

Interviews and the data collected were used to understand how changes and investments in the universities' technology-transfer policies and organization influenced their ability to commercialize their academic ideas.

## DATA COLLECTION

To gain insight into the development of the two regions, Cambridge and New Haven, information was collected from interviews, documents, reports, and secondary sources. Data were collected on several levels of the respective regional economies: the local university, local- and state-level policy documents, industry associations, and relevant companies. From the universities, I gathered mission statements, licensing and patenting policies, faculty promotion policies, general culture with regard to university-industry relationships and how they are reflected in the universities' different departments, as well as technology-transfer policies and organization. From companies, I gathered information on date of foundation, location, the genesis of the knowledge base or technologies, reasons for locating in the specific region, source of employees, relationships with local suppliers and research laboratories (including university research labs), demand conditions, links and services from the local university, and inter-industry linkages.

*Interviews*

The main method of data collection in this study was face-to-face interviews, between sixty and ninety minutes long. The study is based on 115 in-depth interviews, conducted with university administrators on the directorial and managerial levels, including administrators in the highest level of the technology-transfer organization and current and past employees. Interviews were conducted with academics at each university, including representatives from different departments and with different approaches to the technology-transfer process. Senior management, including chief executive officers and presidents of companies, as well as researchers from biotechnology and local pharmaceutical companies, were interviewed. Local trade associations and government representatives, as well as city and state or national representatives, are included under "other organizations" in Table A2.1, which describes the distribution of interviews. The sample includes firms with various ties to their local university—firms that license technology from the university, employ the university's graduates, work in tandem with university research labs, or are companies founded by university graduates or researchers; it also includes some firms that do not work with the university.

TABLE A2.1 Interview distribution

| Region | Academics | Serial entrepreneurs and venture capitalists | Companies | University administrators | Other organizations | Total |
|---|---|---|---|---|---|---|
| Cambridgeshire | 14 | 14 | 26 | 9 | 6 | 69 |
| New Haven | 14 | 2 | 21 | 6 | 3 | 46 |

NOTE: Interviews included firms with and without ties to their local university.

The differences between the two universities and the two biotechnology clusters is reflected in the number of interviews conducted. The Cambridgeshire biotechnology cluster is larger than the New Haven cluster by number of companies (109 companies versus 49). Moreover, technology transfer at the University of Cambridge is dispersed among many units, whereas Yale has just one technology-transfer office. Hence, sixty-nine interviews were conducted in Cambridgeshire County, compared with forty-six interviews in New Haven County.

The main focus of the interviews was university academics. as well as biotechnology and pharmaceutical companies. The interviews provided an extensive understanding of the networking and communications between academia and industry. In addition, interviews were conducted with the university administration that provides and enables these connections. As can be seen in Tables A2.1, A2.2, and A2.3, there are nine interviews with Cambridge administrators and only five with Yale administrators. The reason for the difference is that the University of Cambridge's technology-transfer organization is complex and includes many different units, whereas Yale University has only one technology-transfer office. Another important difference between the two sets of interviews is in the number of serial entrepreneurs and venture capitalists. The reason for this difference is that many of the national and international venture capital firms have a branch in the city of Cambridge but not in New Haven. This fact is meaningful for the development of the biotechnology cluster in each region. My research attempted to understand what these professionals thought of the cluster and the university. As a result, I conducted fourteen interviews with serial entrepreneurs and venture capitalists in Cambridgeshire and two interviews in New Haven.

Most of the interviewees spoke under condition of anonymity. In small- to medium-size companies, providing the title of the interviewee with the name of the company did not allow for confidentiality. Hence, while striving for research openness, and to respect the wishes of the interviewees, the lists of interviews (Tables A2.2 and A2.3) provide only the titles and not the names of the particular companies and organizations. The result of the anonymity was candid interviews that provided much-needed information. Interviews were recorded, transcribed, and analyzed.

To analyze the data, the same questions were repeated to all interviewees in both case studies. University administrators were asked about the university organization, regional commitment, economic development, and technology-transfer commitment. Technology-transfer officers were

TABLE A2.2 List of interviewees at Yale

ACADEMIC FACULTY

Therapeutic radiology

Pharmacology

Molecular, cellular, and developmental biology

Genetics and cellular and molecular physiology

Psychiatry and neurosurgery

Psychiatry and neurosurgery

Applied physics

Genetics

Molecular, cellular, and developmental biology

Therapeutic radiology

Chemical and biomedical engineering

Pharmacology

Cell biology and immunobiology

Molecular biophysics and biochemistry

ENTREPRENEUR, VC, AND INDUSTRY REPRESENTATIVES

Venture capital company, chief executive officer (CEO) and serial entrepreneur

State venture capital company, biotechnology director

BIOTECHNOLOGY AND PHARMACEUTICAL COMPANIES

Chief financial officer (CFO)

CSO

Director of operations

Vice president, biology

Founder and CEO

Director of finance and business development

CFO

Senior director, R&D

Director of microbiology

Associate director, bioanalytical development

Founder and CEO

President and CEO

Researcher

Founder and CFO

President and CEO

President and CEO

CEO

Vice president, collaborative research

Pharmaceutical company, senior vice president, research

Vice president, business development

CEO

Technology-transfer office, director

Technology-transfer office, former director, former academic, and former pharmaceutical executive

University administration, Unit A, director

University administration, vice president

University administration, Unit C, director

University administrator

State industry office, employee

Industry association, director

State industry office, director

NOTE: To provide anonymity, the list of interviewees includes only titles and affiliation description.

TABLE A2.3 List of interviewees at Cambridge

Biotechnology

Chemistry

Blood centre

Pharmacology

Chemistry

Chemistry

Engineering

Laboratory of molecular biology

Pharmacology

Brain repair, Addenbrooke's

Pharmacology

Pathology

(*continued*)

Genetics

Pathology

**ENTREPRENEUR, VC, AND INDUSTRY REPRESENTATIVES**

Director

Serial entrepreneur

Principal, science and technology

Chief executive officer (CEO)

Director

Executive director

Finance industry, life science specialist

Serial entrepreneur

Associate director, technology group

Head of European group

Managing director

Advisor

Non-executive director

Serial entrepreneur

**BIOTECHNOLOGY AND PHARMACEUTICAL COMPANIES**

CEO

Chief scientific officer

Executive director

Cofounder and CEO

Cofounder and CEO

CEO

CEO

CEO

CEO

CEO

CEO

Director

Senior vice president, preclinical R&D

Director

Chairman

Acting CEO

Managing director

CEO

Business development manager

CEO

Chief scientific officer

Chairman

CFO

Cofounder, president and CEO

Founder, scientific advisory board

Business development and marketing executive

UNIVERSITY REPRESENTATIVES

TTO director

TTO, biotechnology manager

Project manager

CEO

Director, entrepreneurship

Director of research policy

Liaison officer

Corporate partnership manager

Director of research policy

OTHER ORGANIZATIONS

Industry association, CEO

Project manager

Industry representative, business development

Regional agency, life science director

Director

Project manager

NOTE: To provide anonymity, the list of interviewees includes only titles and affiliation description.

asked about intellectual property policy, size, and professionalism of their organization, as well as their relationship with the region and the industry. Companies were asked to provide the source of their technology, employees, and advisory boards. They were asked to provide their opinion on the university's academic expertise, technology-transfer accessibility, and professionalism, as well as general questions on the cluster as a whole and the region in which they operate. This method of analysis provided a deep understanding of the two case studies, the way they operate, and their similarities and differences.

To understand the choices of interviews within the universities, organizational charts are provided in Appendix 1. There is much difference in the way each university's organizational chart was retrieved. The University of Cambridge does not have its own organizational chart; it took repeated emails and interviews to the different technology-transfer units at the university, the communication office, and several other administrative offices at the university to gather some information. The communication office at Cambridge finally assisted me with comprehending this information by sending an article from the University of Cambridge's student magazine, *Varsity*. The article, titled "How Does Cambridge Really Work?" (Lester 2007) provided a simple organizational chart of the university that did not include the technology-transfer units. I added these by using information received directly from each technology-transfer unit at the university. The organizational chart of the University of Cambridge marks the location of technology transfer in the organization. The chart, while not as detailed as the Yale chart, provides a clear description of the multiple technology-transfer units at the university. Yale University's organizational chart was easily retrieved from a simple web search and few inquiries at the university. The chart has much detail, including the technology-transfer functions at the university. Yale University has one technology-transfer unit, which is clearly marked on the organizational chart of the university.

### Secondary Resources

This study examined annual reports and websites of universities, industry firms, trade associations, and state and county governments. Other data were gathered from articles and books on national and regional innovation policy as well as from the universities and clusters. The information obtained from these sources allowed a comparison of the findings with those of other cases, to come to several conclusions regarding how changes and investments in technology transfer influence the ability of universities to commercialize their academic ideas and collaborate with local industry.

CHAPTER ONE

1. Biotechnology can be defined as "any technique that uses living organisms, or substances from those organisms, to make or modify a product, to improve plants or animals, or to develop microorganisms for specific uses" (Barnum 1998).

2. Biotechnology companies take about fifteen years to get to the proof of concept. During this time, many choose to stay in close contact with university and research labs.

3. Except Oxford, which has a similar model to that of Cambridge.

4. The University of Cambridge did not register all of its patents at both the European Patent Office and the US Patent Office. The largest number of patents registered to the university is at the European Patent Office, and those are the numbers reflected in this research.

5. As a measure of comparison, federal government funding at MIT is 50 percent, and industry provides 3 percent of total operating revenues.

6. "Sciences" here includes physical sciences, technology, and biological sciences (University of Cambridge 2013).

7. At Cambridge, the number of medical school faculty members is included within the sciences faculty (University of Cambridge 2013; Massachusetts Institute of Technology 2013; Yale University 2011–12).

8. Professor Frank Ruddle developed the transgenic mouse at Yale in 1980. He was one of the first scientists in the world to do so. However, Yale University never patented the invention.

9. One invention that was patented during this period was the profitable drug Zerit. This drug, which was licensed to Bristol-Myers Squibb, is part of the AIDS drug combination treatment.

CHAPTER TWO

1. The Department of Trade and Industry is now the Department for Business, Innovation, and Skills.

2. The State of Connecticut created Connecticut Innovations in 1989 to provide seed funding for companies in Connecticut. By 1995, Connecticut Innovations had returned its investment to the state, and it is now an independent investment company.

3. More about the impact of the different individuals in Cambridge and Yale can be found in Chapters 4 and 5. Bruce Alexander is a prominent real estate and management executive. Some of the projects developed under his directions were Harborplace in Baltimore, Miami's Bayside, and Portland's Pioneer Place.

4. More about the economic development program initiated by President Levin can be found in Chapter 4.

5. A detailed explanation on government reports and funding that affected technology transfer at Cambridge can be found in Chapter 5.

6. For more about how industry views the offices, see Chapters 4 and 5.

7. Due diligence is "the investigation of an asset, investment, or anything else to ensure that everything is as it seems. Due diligence helps a buyer or investor make sure that there are no unexpected problems with the asset or investment and that he/she does not overpay. Due diligence can be a complex and formalized process in the acquisition of a company. Even when buying a house, for example, due diligence involves time consuming and at times expensive endeavors, like a home inspection. However, due diligence is seen as a necessary part of doing business or buying an asset" (Farlex Financial Dictionary 2012).

8. Although Cambridge University does not have a big success in biotech, it has shown success in IT with firms such as ARM (http://www.arm.com).

## CHAPTER THREE

1. At different times money was appropriated through legislation such as the second Morrill Act and the Bankhead-Jones Act, although the funding provisions of these acts are no longer in effect. Today, the Nelson Amendment to the Morrill Act provides a permanent annual appropriation of $50,000 per state and territory.

2. The Smith-Lever Act was directed mainly to the agricultural experiment stations.

3. "The SBIR program was established under the Small Business Innovation Development Act of 1982 (P.L. 97-219) with the purpose of strengthening the role of innovative small business concerns in Federally-funded research and development (R&D). Through FY2009, over 112,500 awards have been made totaling more than $26.9 billion. In December 2000, Congress passed the Small Business Research and Development Enhancement Act (P.L. 102-564), reauthorizing the SBIR program until September 30, 2000. The program was reauthorized until September 30, 2008 by the Small Business Reauthorization Act of 2000 (P.L. 106-554). Subsequently, Congress passed numerous extensions, the most recent of which extends the SBIR program through 2017" (Small Business Administration 2013).

4. *Universities* refers to both public and private. All funding (e.g., NIH, NSF) was received from government bodies.

5. Patent potential means that the invention is new or additional technology, with the potential to bring financial revenues to the university.

6. The UK higher education institutions' funding is based on several sources: (1) the funding council grants, including those from the Higher Education Funding Council for England, the Higher Education Funding Council for Wales, the Scottish Further and Higher Education Funding Council, the Training and Development Agency for Schools, and the Department for Employment and Learning–Northern Ireland; (2) tuition fees and education grants and contracts; (3) research grants and contracts, including research councils covered by the Office of Science and Technology, UK-based charities, UK central government bodies, local authorities, health and hospital authorities, UK industry, commerce and public corporations, EU government bodies, and other overseas sources, including all research grants and contracts income from overseas bodies operating outside the European Union; and (4) other income, which includes residences and catering operations, grants from local authorities, income from health and hospital authorities, release of deferred capital grants, income from intellectual property rights, and other operating income.

## CHAPTER FOUR

1. A bridge loan is a short-term loan used until a person or company can arrange more comprehensive long-term financing. The need for a bridge loan arises when a company runs out of cash before it can obtain more capital investment through long-term debt or equity. Series A preferred stock is the first round of stock offered during the seed or early-stage round by a portfolio company to the venture capitalist. Series A preferred stock is convertible into common stock in certain cases, such as an initial public offering or the sale of the company. Later rounds of preferred stock in a private company are called Series B, Series C, and so on.

2. Connecticut Innovations invests in many industrial sectors that show potential for the Connecticut economy. These include bioscience, information technology, energy and environmental systems, photonics, and others.

3. The duties of the OCR include oversight for patenting and licensing activities, university inventions, and contractual relationships between faculty and industry. The staff of the OCR work with Yale researchers to identify inventions that may ultimately become commercial products and services useful to the public. Staff also engage in industrial partnerships to license Yale inventions. An important goal for the Yale OCR is to identify new ideas, cultivate venture funding for them, and facilitate their development into companies that become part of the New Haven economy (OCR 2003).

4. Stamford and Greenwich are two towns in Fairfield County in Connecticut, on the New York border. In the 1980s many corporations, including financiers, moved from New York both to lower their tax bills and to be closer to the homes of their top executives, who chose to build their houses outside New York City. Thus, Connecticut has a large concentration of venture capitalists living in the New Haven metropolitan area (which includes New Haven and Fairfield Counties).

5. Yale's ability to recruit top talent is demonstrated in the recruitment of both Greg Gardiner, from Pfizer, and Bruce Alexander. This also confirms that Yale had a choice in who to recruit and when to recruit them.

6. Based on an interview with a Yale faculty member.

7. Frank Ruddle invented the transgenic mouse, and Tommy Cheng invented several drugs to treat cancer, HIV/AIDS, and hepatitis B.

8. The School of Medicine includes the Department of Medicine, the Department of Epidemiology and Public Health, and the Physician Associate program.

9. In total, Yale has spun out forty-three biotechnology companies as of 2012, of which twenty-seven chose to locate in the New Haven area.

10. As an example, we can see that the number of Yale's spinouts in the cluster grew by 50 percent from 2000 to 2004.

11. Although one person's view not is not statistically significant, it does tell us something about how important biotechnology is to the New Haven economy.

12. Connecticut Innovations is the state's leading investor in high technology.

13. By 2006 Bayer had closed its West Haven facility. Its land was bought by Yale University in 2007 to create a life science campus.

CHAPTER FIVE

1. Colleges in the University of Cambridge are independent institutions, recruiting their own faculty and students, and managing their finances while paying their dues to the university.

2. David Gill replaced Herriot in 2008.

3. The colleges are autonomous institutes that select their own faculty and students, although they are connected to the university through membership in the university council and representation on the different boards.

4. In 2007 DTI was abolished and the new Office for Business, Innovation, and Skills was created.

5. BioConcepts is an institutional bridge between investors and local firms, first for business angels and seed funds and ultimately for the venture capital community looking for matured early-stage opportunities; i10 supports innovation and success of regional firms by helping companies tap into the expertise and resources of universities in the east of England.

6. A new wave of initiatives that may have a direct impact on the biotechnology industry started in 2010 and 2011. The most noteworthy one is the Strategy for UK Life Sciences, launched in 2011. The strategy includes a creation of the Office of Life Sciences under the Office of Business, Innovation, and Skills, which replaced DTI in 2007. Through the strategy £310 million was allocated to investment in research and commercialization, with £180 million for the Biomedical Catalyst Fund, which provides funding for small and medium-size enterprises and academic institutions in the health-care field. However, the results of this initiative are too early to evaluate here.

7. The purchase of shares and the pressure to choose which company CSI had to merge with was considered unprecedented, violating the government's own policy of non-involvement.

8. The *Cambridge University Reporter* is the official journal of the University of Cambridge. It carries notices of all university business, announcing university events, proposals for changes in regulations, council and general board decisions,

as well as information on awards, scholarships, and appointments at Cambridge and other universities. As such, the *Reporter* is not only a newspaper but also the official channel of communication and information delivery between the university and its employees, students, and faculty.

9. Cambridge Enterprise replaced WILO in 1999. The description of the technology-transfer organization is in the next section.

10. The University of Cambridge did not register all of its patents at both the European Patent Office and the US Patent Office. The largest number of patents registered to the university is at the European Patent Office, and those are the numbers reflected in this research.

11. Unlike Yale, which did not have a biotechnology cluster and had to create policies to encourage the creation of spinout companies in the region.

12. The data analysis from VentureXpert could not find information for about 40 percent of the firms, and hence they may have received funding from other sources.

13. These numbers reflect total number of companies, not spin-outs.

14. For more about industry response, see the following section for a discussion on the industry view of Cambridge University.

15. An official position means an active role in the company and not just on the scientific advisory board.

16. Note that the university may have attempted to coordinate research collaborations with industry because of the number and consequences of contracts that were loss making because of a pervasive and consistent underestimate of overhead costs. However, there is no evidence to support this line of reasoning.

17. The changes to Cambridge enterprise and research services division continue. By 2013 RSD had become the Research Operations Office, and it includes mainly research grants and contracts. The liaison program moved back into Cambridge Enterprise.

## CHAPTER SIX

1. "The Patent Board is a business-based patent adviser to Fortune 500 companies, technology-based start-ups, law firms, investment banks, and governments. The Patent Board uses proprietary data, tools, and analytics to leverage patent-based intellectual property as an asset class. The Patent Board tracks and analyzes innovation, movement, and the business impact of patent assets across seventeen industries on a global basis" (Oldach 2009).

## CHAPTER SEVEN

1. Related activities refer to entrepreneurship education, invention competitions, and the like.

2. Some incubators and science parks are owned or partially owned by a university.

# References

Acharya, R. 1999. *The emergence and growth of biotechnology.* Cheltenham, UK: Edward Elgar.

Adams, J., and B. Bekhradnia. 2004. *What future for dual support?* Oxford, UK: Higher Education Policy Institute.

Amin, A., N. J. Thrift, and ESF Programme on Regional and Urban Restructuring in Europe. 1994. *Globalization, institutions, and regional development in Europe.* Oxford: Oxford University Press.

Association of University Technology Managers (AUTM). 2009. *AUTM licensing activity surveys.*

———. 2013. *Statistics access for tech transfer (STATT).* http://www.autmsurvey.org/statt/search.cfm?login=true&CFID=80627098&CFTOKEN=89444475.

Atlas, S. 1996. Yale student's killer convicted once again. *Yale Daily News.*

Audretsch, D. B., J. Weigand, and C. Weigand. 2002. The impact of the SBIR on creating entrepreneurial behavior. *Economic Development Quarterly* 16 (1): 32–38.

Bagchi-Sen, S., and H. Lawton Smith. 2012. The role of the university as an agent of regional economic development. *Geography Compass* 6 (7): 439–53.

Ball, M. 1999. Is Yale safe? *Yale Herald,* October 15.

Bandura, A. 1977. *Social learning theory.* Englewood Cliffs, NJ: Prentice-Hall.

Barnum, S. R. 1998. *Biotechnology: An introduction.* Belmont, CA: Wadsworth Publishing.

Bartik, T. J. 1985. Business location decisions in the United States—Estimates of the effects of unionization, taxes, and other characteristics of states. *Journal of Business and Economic Statistics* 3 (1): 14–22.

Beck, R., D. Elliott, J. Meisel, and M. Wagner. 1995. Economic impact studies of regional public colleges and universities. *Growth and Change* 26 (2): 245–60.

Bercovitz, J., and M. Feldman. 2005. Academic entrepreneurs: Organizational change at the individual level. Paper presented at Technology Transfer Society, Kansas City, MO, September 28–30.

——. 2007. Fishing upstream: Firm innovation strategy and university research alliances. *Research Policy* 36 (7): 930–48.

——. 2008. Academic entrepreneurs: Organizational change at the individual level. *Organization Science* 19 (1): 69–89.

Bercovitz, J., M. Feldman, I. Feller, and R. Burton. 2001. Organizational structure as a determinant of academic patent and licensing behavior: An exploratory study of Duke, Johns Hopkins, and Pennsylvania State Universities. *Journal of Technology Transfer* 26 (1): 21–35.

Blankenburg, S. 1998. *University-industry relations, innovation and power: A theoretical framework for the study of technology transfer from the science base.* Cambridge: Economic and Social Research Council, Center for Business Research, University of Cambridge.

Blumenstyk, G. 1990. Yale agrees to pay New Haven for some city services. *Chronicle of Higher Education*, April 11, A30.

——. 2001. Turning patent royalties into a sure thing. *Chronicle of Higher Education*, October 5.

Blumenthal, D., N. Causino, E. Campbell, and K. Seashore Louis. 1996. Relationships between academic institutions and industry in the life sciences: An industry survey. *New England Journal of Medicine* 334 (6): 368–74.

Boucher, G., C. Conway, and E. Van Der Meer. 2003. Tiers of engagement by universities in their region's development. *Regional Studies* 37 (9): 887–97.

Breznitz, S. M. 2000. The geography of industrial districts: Why does the biotechnology industry in Massachusetts cluster in Cambridge? MA thesis, Regional Economic and Social Development, University of Massachusetts, Lowell.

——. 2007. From ivory tower to industrial promotion: The development of the biotechnology cluster in New Haven, Connecticut. *Revue d'Economie Industrielle* 120 (4): 115–34.

——. 2011. Improving or impairing? Following technology transfer changes at the University of Cambridge. *Regional Studies* 45 (4): 463–78.

Breznitz, S. M., and W. Anderson. 2006. Boston metropolitan area biotechnology cluster. *Canadian Journal of Regional Science* 28 (2): 249–64.

Breznitz, S. M., and M. P. Feldman. 2012. The larger role of the university in economic development. *Journal of Technology Transfer* 37 (2): 135–38.

Breznitz, S. M., R. P. O'Shea, and T. J. Allen. 2008. University commercialization strategies in the development of regional bioclusters. *Journal of Product Innovation Management* 25 (3): 129–42.

Breznitz, S. M., and N. Ram. 2013. Enhancing economic growth? University technology commercialization. In *Creating competitiveness*, edited by D. B. Audretsch and M. L. Walshok, 88–115. Cheltenham, UK: Edward Elgar.

Cambridge Enterprise. 2009. *About us.* 2009. http://www.enterprise.cam.ac.uk/company-information/.

——. 2013a. *Company information—People.* http://www.enterprise.cam.ac.uk/company-information/people/.

——. 2013b. *Mission-goals-principles.* http://www.enterprise.cam.ac.uk/company-information/mission-goals-principles/.

*Cambridge University Reporter.* 2001. *Ownership of intellectual property rights generated by externally funded research: Notice (5835).* http://www.admin.cam.ac.uk/reporter/.

*Cambridge University Reporter.* 2005. Third joint report of the Council and the General Board on the ownership of intellectual property rights (IPRs): Notice (6008). http://www.admin.cam.ac.uk/reporter/.

Cantwell, J. 1987. The reorganization of European industries after integration: Selected evidence on the role of multinational enterprise activities. *Journal of Common Market Studies* 26 (2): 127–51.

Carlsson, B., and A.-C. Fridh. 2002. Technology transfer in United States universities. *Journal of Evolutionary Economics* 12 (1): 199–232.

Carlsson, B., S. Jacobsson, M. Holmen, and A. Rickne. 2002. Innovation systems: Analytical and methodological issues. *Research Policy* 31 (2): 233–45.

Casper, S. 2007. How do technology clusters emerge and become sustainable? Social network formation and inter-firm mobility within the San Diego biotechnology cluster. *Research Policy* 36 (4): 438–55.

Chapple, W., A. Lockett, D. Siegel, and M. Wright. 2005. Assessing the relative performance of UK university technology transfer offices: Parametric and non-parametric evidence. *Research Policy* 34 (3): 369–84.

Clark, B. R. 1998. *Creating entrepreneurial universities: Organizational pathways of transformation.* Oxford, UK: Pergamon Press.

Clarysse B., M. Wright, A. Lockett, E. van de Elde, and A. Vohora. 2005. Spinning out new ventures: A typology of incubation strategies from European research institutions. *Journal of Business Venturing* 20: 183–216.

Committee on Higher Education Under the Chairmanship of Professor Nevill Mott. 1967. *Higher education report.* Cambridge: University of Cambridge.

Connecticut Economic Development. 2013. *Life sciences/medical devices.* http://cteconomicdevelopment.com/CT-biotech-companies.php.

Connecticut Innovations. 2003. *About.* http://www.ctinnovations.com/about.

Connecticut United for Research Excellence (CURE). 2003. *Economic Survey Report.* New Haven, CT: CURE.

Cooke, P. 2002. Regional innovation systems: General findings and some new evidence from biotechnology clusters. *Journal of Technology Transfer* 27 (1): 133–45.

Cooke, P., and K. Morgan. 1998. *The associational economy: Firms, regions, and innovation.* Oxford: Oxford University Press.

Dacin, M. T., J. Goodstein, and W. R. Scott. 2002. Institutional theory and institutional change: Introduction to special research forum. *Academy of Management Journal* 45 (1): 45–57.

Dearing, R. 1997. *National Committee of Inquiry into Higher Education.* https://bei.leeds.ac.uk/Partners/NCIHE/.

Department of Chemistry, University of Cambridge. 2006. *Corporate associates scheme.* http://www-cas.ch.cam.ac.uk/about/.

Department of Trade and Industry (DTI). 2005. *Higher education institutions—Welcome.* http://www.ktponline.org.uk/hei/.

———. 2006. *DTI innovation policy*. http://www.dti.gov.uk/innovation/innovation-dti/page11863.html.

Di Gregorio, D., and S. Shane. 2003. Why do some universities generate more start-ups than others? *Research Policy* 32: 209–27.

DiMaggio, P. J. 1988. Interest and agency in institutional theory. In *Institutional patterns and organizations: Culture and environment*, edited by L. G. Zucker, 3–23. Cambridge, MA: Ballinger.

DiMaggio, P., and W. W. Powell. 1983. The iron cage revisited: Institutional isomorphism and collective rationality in organizational fields. *American Sociological Review* 48 (2): 147–60.

Druilhe, C., and E. Garnsey. 2004. Do academic spin-outs differ and does it matter? *Journal of Technology Transfer* 29 (3–4): 269–85.

Dun and Bradstreet. 2005. *D&B million dollar database*.

East of England Development Agency. 2005a. *Venture capital fund 2005*. http://www.eeda.org.uk/abouteeda/tour.html.

———. 2005b. *Who we are 2005*. http://www.eeda.org.uk/abouteeda/tour.html.

Eastern Region Biotechnology Initiative. 2004. *Biotechnology in Cambridge and the East of England*. http://www.onenucleus.com.

———. 2005–6. *Directory*. http://www.onenucleus.com.

Eaton, S. C., and L. Bailyn. 1999. Work and life strategies of professionals in biotechnology firms. *Annals of American Academy of Political and Social Science* 562: 159–73.

Ernst & Young. 2001. *Focus on fundamentals: The biotechnology report*.

Etzkowitz, H. 1995. The triple helix—University-industry-government relations: A laboratory for knowledge based economic development. *EASST Review* 14: 9–14.

———. 1998. The norms of entrepreneurial science: Cognitive effects of the new university-industry linkages. *Research Policy* 27 (8): 823–33.

———. 2002. *MIT and the rise of entrepreneurial science*. London: Routledge.

Etzkowitz, H., and L. Leydesdorff. 2000. The dynamics of innovation: From national systems and "Mode 2" to a triple helix of university-industry-government relations. *Research Policy* 29 (2): 109–23.

Etzkowitz, H., A. Webster, C. Gebhardt, and B. Regina Cantisano Terra. 2000. The future of the university and the university of the future: Evolution of ivory tower to entrepreneurial paradigm. *Research Policy* 29 (2): 313–30.

European Patent Office. 2006. *Patents published*. http://www.epo.org.

———. 2013. *Patents published*. http://www.epo.org.

Farlex Financial Dictionary. 2012. *Due diligence*.

Federation of Finnish Technology Industries. 2010. *Turnover remains extremely low, staffing changes continue*. http://www.teknologiateollisuus.fi/en/news/announcements/2010-2/turnover-remains-extremely-low-staffing-changes-continue.

Feldman, M. P., and S. M. Breznitz. 2009. The American experience in university technology transfer. In *European universities learning to compete: From social institutions to knowledge business*, edited by M. McKelvey and M. Holmén, 161–86. Cheltenham, UK: Edward Elgar.

Feldman, M. P., and P. Desrochers. 2003. Research universities and local economic development: Lessons from the history of Johns Hopkins University. *Industry and Innovation* 10: 5–24.

Feldman, M. P., I. Feller, J. E. L. Bercovitz, and R. M. Burton. 2002. Equity and the technology transfer strategies of American research universities. *Management Science* 48: 105–21.

Felsenstein, D. 1996. The university in the metropolitan arena: Impacts and public policy implications. *Urban Studies* 33 (9): 1565–80.

Foltz, J., B. Barham, and K. Kim. 2000. Universities and agricultural biotechnology patent production. *Agribusiness* 16 (1): 82–95.

Freeman, C. 1995. The national innovation systems in historical perspective. *Cambridge Journal of Economics* 19 (1): 5–24.

———. 2002. Continental, national and sub-national innovation systems: Complementarity and economic growth. *Research Policy* 31 (2): 191–211.

Garnsey, E., and P. Hefferman. 2005. High-technology clustering through spin-out and attraction: The Cambridge case. *Regional Studies* 39 (8): 1127–44.

Geuna, A., and L. J. J. Nesta. 2006. University patenting and its effects on academic research: The emerging European evidence. *Research Policy* 35 (6): 790–807.

Goldstein, H. A., and C. S. Renault. 2004. Contributions of universities to regional economic development: A quasi-experimental approach. *Regional Studies* 38 (7): 733–46.

Gray, M., and S. Damery. 2003. *Regional development and differentiated labor markets: The Cambridge case.* Report to the European Commission, IST Program.

Great Britain House of Commons. 1963. *Higher education: Government statement on the report of the committee under the chairmanship of Lord Robbins, 1961–63.* London: Her Majesty's Stationery Office.

Greater Cambridge Greater Peterborough Enterprise Partnership. 2013. *About the Enterprise Partnership* 2013. http://www.yourlocalenterprisepartnership .co.uk.

Greenwood, R., and C. R. Hinings. 2006. Radical organizational change. In *The Sage handbook of organization studies*, 2nd ed., edited by C. Hardy, S. R. Clegg, W. Nord, and T. Lawrence, 814–42. London: Sage.

Harrison, B., and M. Weiss. 1998. *Workforce development networks.* Thousand Oaks, CA: Sage.

Hatakenaka, S. 2002. Flux and flexibility: A comparative institutional analysis of evolving university-industry relationships in MIT, Cambridge and Tokyo. PhD diss., Sloan School of Management, Massachusetts Institute of Technology.

Higher Education Funding Council for England. 2003. *Higher education-business interaction survey 2000–01.* http://webarchive.nationalarchives.gov .uk/20100202100434/http://hefce.ac.uk/pubs/hefce/2003/03_11.htm.

HM Revenue and Customs. 2013. *Research and development (R&D) relief for corporation tax.* http://www.hmrc.gov.uk/ct/forms-rates/claims/randd.htm#6.

Jaffe, A. B., M. Trajtenberg, and R. Henderson. 1993. Geographic localization of knowledge spillovers as evidenced by patent citations. *Quarterly Journal of Economics* 108 (3): 577–98.

James, A. 2005. Demystifying the role of culture in innovative regional economies. *Regional Studies* 39 (9): 1197–1216.

Jennings, R. 1994. Why the university's technology transfer company is called "Lynxvale LTD." *Magazine of Cambridge Society* 35.

Jensen, R. A., J. G. Thursby, and M. C. Thursby. 2003. Disclosure and licensing of university inventions: "The best we can do with the s—t we get to work with." *International Journal of Industrial Organization* 21 (9): 1271–1300.

Keeble, D. 2001. *University and technology: Science and technology parks in the Cambridge region*. Cambridge: Center for Business Research, University of Cambridge.

Kenney, M. 1986. *Biotechnology: The university-industrial complex*. New Haven, CT: Yale University Press.

Kenney, M., and R. W. Goe. 2004. The role of social embeddedness in professional entrepreneurship: A comparison of electrical engineering and computer science at UC Berkeley and Stanford. *Research Policy* 33: 691–707.

Kenney, M., and D. Patton. 2009. Reconsidering the Bayh-Dole Act and the current university invention ownership model. *Research Policy* 38: 1407–22.

Koepp, R. 2002. *Clusters of creativity: Enduring lessons on innovation and entrepreneurship from Silicon Valley and Europe's Silicon Fen*. Chichester, UK: Wiley.

Kortum, S., and J. Lerner. 2000. Assessing the contribution of venture capital to innovation. *RAND Journal of Economics* 31: 674–92.

Krimsky, S. 1987. The new corporate identity of the American university. *Alternatives* 14 (2): 20–29.

———. 1988. University entrepreneurship and the public purpose. *Biotechnology: Professional Issues and Social Concerns* 88 (23): 34–42.

Kunda, G. 1992. *Engineering culture: Control and commitment in a high-tech corporation* Philadelphia: Temple University Press.

Lambert, R. 2003. *Lambert review of business-university collaboration*. http://www.eua.be/eua/jsp/en/upload/lambert_review_final_450.1151581102387.pdf.

Lane, J. R., and B. D. Johnstone, eds. 2012. *Universities and colleges as economic drivers*. Albany: State University of New York Press.

Lawson, C., and E. Lorenz. 1999. Collective learning, tacit knowledge and regional innovative capacity. *Regional Studies* 33 (4): 305–17.

Lawton Smith, H. 2006. *Universities, innovation, and the economy*. Abington, UK: Routledge.

Lawton Smith, H., and S. Bagchi-Sen. 2012. The research university, entrepreneurship and regional development: Research propositions and current evidence. *Entrepreneurship and Regional Development* 24 (5–6): 383–404.

Lawton Smith, H., and K. Ho. 2006. Measuring the performance of Oxford University, Oxford Brookes University and the government laboratories' spin-off companies. *Research Policy* 35 (10): 1554–68.

Lester, R. 2007. How does Cambridge really work? *Varsity*, February 16, 6–7.

Levin, R. C. 1993. *Beyond the ivy walls: Our university in the wider world*. Yale Office of the President. October 2. http://www.yale.edu/opa/president/speeches/levin_inaugural.html.

———. 2003. *Universities and cities: The View from New Haven.* Yale Office of the President. January 30. http://www.yale.edu/opa/president/speeches/case _western_20030130.html.

Link, A. N., and J. T. Scott. 2005. Opening the ivory tower's door: An analysis of the determinants of the formation of US university spin-off companies. *Research Policy* 34 (7): 1106–12.

Link, A., and D. Siegel. 2005. Generating science-based growth: An econometric analysis of the impact of organizational incentives on university-industry technology transfer. *European Journal of Finance* 11 (3): 169–81.

Litan, R. E., and L. Mitchell. 2010. A faster path from lab to market. *Harvard Business Review* (January–February): 52–53.

Litan, R. E., L. Mitchell, and E. J. Reedy. 2007. The university as innovator: Bumps in the road. *Issues in Science and Technology* 23 (4): 57–66.

Lockett, A., and M. Wright. 2005. Resources, capabilities, risk capital and the creation of university spin-out companies. *Research Policy* 34 (7): 1043–57.

Lundvall, B.-A. 1994. The learning economy: Challenges to economic theory and policy. Paper presented at the European Association for Evolutionary Political Economy conference, Copenhagen.

Lundvall, B.-A., B. Johnson, E. Sloth Andersen, and B. Dalum. 2002. National systems of production, innovation and competence building. *Research Policy* 31 (2): 213–31.

Markusen, A. R. 1991. *The rise of the gunbelt: The military remapping of industrial America.* New York: Oxford University Press.

Mass Biotechnology Cluster. 2013. *Incentives.* http://www.massbio.org/ economic_development/massachusetts_incentives.

Matheson, D., and M. B. Silverstein. 2002. *MassBiotech 2010: Achieving global leadership in the life-sciences economy.* http://www.forskningsradet.no/CSStorage/ Flex_attachment/BiotekBCGMassBiotech.pdf.

Massachusetts Institute of Technology. 2013. *MIT facts.* http://web.mit.edu/facts/ faculty.html.

Meyer, J. W., and B. Rowan. 1977. Institutionalized organizations: Formal structure as myth and ceremony. *American Journal of Sociology* 83 (2): 440–63.

Miner, A. S., D. T. Eesley, M. Devaughn, and T. Rura-Polley. 2001. The magic beanstalk vision: Commercializing university inventions and research. In *Entrepreneurial dynamic*, edited by C. B. Schoonhoven and E. Romanelli, 109–46. Stanford, CA: Stanford University Press.

Minshall, T., C. Druilhe, and D. Probert. 2004. The evolution of "third mission" activities at the university of Cambridge: Balancing strategic and operational considerations. Presentation at the 12th High Tech Small Firms Conference, University of Twente, The Netherlands. May 24–25.

Minshall, T., and B. Wicksteed. 2005. *University spin-out companies? Starting to fill the evidence gap.* Cambridge, UK: Gatsby Charitable Foundation.

Morgan, K. 1997. The learning region: Institutions, innovation and regional renewal. *Regional Studies* 31 (5): 491–503.

Moulaert, F., and F. Sekia. 2003. Territorial innovation models: A critical survey. *Regional Studies* 37 (3): 289–302.

Mowery, D. C., R. R. Nelson, B. N. Sampat, and A. A. Ziedonis. 2004. *Ivory tower and industrial innovation: University-industry technology transfer before and after the Bayh-Dole Act*. Stanford, CA: Stanford University Press.

Mowery, D., R. R. Rosenberg, B. N. Sampat, and A. A. Ziedonis. 1999. The effects of the Bayh-Dole Act on US university research and technology transfer. In *Industrializing knowledge: University-industry linkages in Japan and the United States*, edited by L. M. Branscomb, F. Kodama, and R. L. Florida, 269–306. Cambridge, MA: MIT Press.

Mowery, D. C., and B. N. Sampat. 2001a. Patenting and licensing university inventions: Lessons from the history of the research corporation. *Industrial and Corporate Change* 10 (2): 317–55.

———. 2001b. University patents and patent policy debates in the USA, 1925–1980. *Industrial and Corporate Change* 10 (3): 781–814.

Murray, F. 2002. Innovation as co-evolution of scientific and technological networks: exploring tissue engineering. *Research Policy* 31 (8–9): 1389–1403.

Myint, Y. M., S. Vyakarnam, and M. J. New. 2005. The effect of social capital in new venture creation: The Cambridge high-technology cluster. *Strategic Change* 14 (May): 165–77.

National Science Foundation. 2003. *Table B-32, Total R&D Expenditures at universities and colleges, ranked by fiscal year 2001*. http://www.nsf.gov/statistics/nsf06329/pdf/tabb32.pdf.

———. 2005. *NSF at a glance*. http://www.nsf.gov/about/glance.jsp.

———. 2013a. *National patterns of R&D resources: 2010–11 data update*. http://www.nsf.gov/statistics/nsf13318/content.cfm?pub_id=4268&id=2.

———. 2013b. *R&D expenditures at universities and colleges, by source of funds FY 1953–2010*. http://www.nsf.gov/statistics/nsf12330/content.cfm?pub_id=4211&id=2.

Nelson, R. R. 1993. *National innovation systems: A comparative analysis*. New York: Oxford University Press.

O'Shea, R. P., T. J. Allen, A. Chevalier, and F. Roche. 2005. Entrepreneurial orientation, technology transfer and spinoff performance of US universities. *Research Policy* 34 (7): 994–1009.

Office of Cooperative Research. 1998. *1996–1998 Yale OCR annual report: "From bench to bedside."* New Haven, CT: Yale University.

———. 2003. *Mission statement*. http://www.yale.edu/ocr/about/index.html.

———. 2012. *Activity report*. New Haven, CT: Yale University.

———. 2013. *Mission statement*. http://www.yale.edu/ocr/about/.

Office of Institutional Research. 2001. *Factsheet*. http://www.yale.edu/oir/factsheet.html.

Oldach, S. 2009. *The Universities Patent Scorecard—2009 university leaders in innovation*. http://www.iptoday.com/articles/2009-9-oldach.asp.

Owen-Smith, J., and W. Powell. 2001. To patent or not: Faculty decisions and institutional success in academic patenting. *Journal of Technology Transfer* 26 (1): 99–114.

———. 2004. Knowledge networks as channels and conduits: The effects of spillovers in the Boston biotechnology community. *Organization Science* 15 (1): 5–21.

Owen-Smith, J., M. Riccaboni, F. Pammolli, and W. W. Powell. 2002. A comparison of US and European university-industry relations in the life sciences. *Management Science* 48 (1): 24–43.

Pfeffer, J. 1978. *Organizational design.* Arlington Heights, IL: AHM Publishing.

Poppick, S. 2005. Former prof loses ruling over patent. *Yale Daily News*, February 14.

Porter, M. E. 1990. *The competitive advantage of nations.* New York: Free Press.

Rahm, D., J. Kirkland, and B. Bozeman. 2000. *University-industry R&D collaboration in the United States, the United Kingdom, and Japan.* Dordrecht, The Netherlands: Kluwer Academic Publishers.

Rappa, J. 2011. *Connecticut's bioscience industry: OLR research report.* Hartford: State of Connecticut General Assembly.

Roberts, E. B. 1991. *Entrepreneurs in high technology: Lessons from MIT and beyond.* New York: Oxford University Press.

Roberts, E. B., and D. E. Malone. 1996. Policies and structures for spinning off new companies from research and development organizations. *R&D Management* 26: 17–48.

Rothaermel, F. T., S. D. Agung, and L. Jiang. 2007. University entrepreneurship: A taxonomy of the literature. *Industrial and Corporate Change* 16 (4): 691–791.

Russell, C. 1993. *Academic freedom.* London: Routledge.

Sainsbury, Lord (Minister for Science). 1999. *Biotechnology clusters.* London: Ministry of Science.

Sampat, B. N. 2006. Patenting and US academic research in the 20th century: The world before and after Bayh-Dole. *Research Policy* 35 (6): 772–89.

Saxenian, A. 1994. *Regional advantage: Culture and competition in Silicon Valley and Route 128.* Cambridge, MA: Harvard University Press.

Schein, E. H. 1985. *Organizational culture and leadership: A dynamic view.* San Francisco: Jossey-Bass.

Schoenberger, E. 1997. *The cultural crisis of the firm.* Cambridge, MA: Blackwell.

Scott, P. 1977. *What future for higher education?* London: Fabian Tracts.

Scott, W. R. 2003. *Organizations: Rational, natural, and open systems.* Upper Saddle River, NJ: Prentice-Hall.

Scott, W. R., and G. F. Davis. 2007. *Organizations and organizing.* Upper Saddle River, NJ: Pearson Prentice Hall.

Sedgwick, J. 1994. The death of Yale. *GQ* 64 (4): 182.

Segal Quince Wicksteed. 1985. *The Cambridge phenomenon: The growth of high technology industry in a university town.* Cambridge, UK: Segal Quince Wicksteed.

———. 2000. *The Cambridge phenomenon revisited.* Histon, UK: Segal Quince Wicksteed.

Shane, S. 2002. Selling university technology: Patterns from MIT. *Management Science* 48 (1): 122–37.

———. 2004. *Academic entrepreneurship: University spinoffs and wealth creation.* Cheltenham, UK: Edward Elgar.

Siegal, D. S., and H. P. Phan. 2005. Analyzing the effectiveness of university technology transfer: Implications for entrepreneurship education. In *University entrepreneurship and technology transfer: Process, design, and intellectual property*, edited by D. G. Libecap, 1–38. Amsterdam: Elsevier.

Siegel, D. S., D. A. Waldman, L. E. Atwater, and A. N. Link. 2004. Toward a model of the effective transfer of scientific knowledge from academicians to practitioners: Qualitative evidence from the commercialization of university technologies. *Journal of Engineering and Technology Management* 21 (1–2): 115–42.

Siegel, D. S., D. Waldman, and A. Link. 2003a. Assessing the impact of organizational practices on the relative productivity of university technology transfer offices: An exploratory study. *Research Policy* 32 (1): 27–48.

———. 2003b. Improving the effectiveness of commercial knowledge transfers from universities to firms. *Journal of High Technology Management Research* 14: 111–33.

Small Business Administration. 2013. *SBIR*. http://www.sbir.gov/about/about-sbir.

Staley, O. 2013. Cambridge enters UK banking business to lift endowment. *Bloomberg*, February 14. http://www.bloomberg.com/news/2013-02-15/cambridge-enters-u-k-banking-business-to-lift-endowment.html.

Statistics Finland. 2010. *Turnover in manufacturing grew by nearly 8 per cent in March to May from the year before.* http://www.stat.fi/org/yhteystiedot/index_en.html.

Steiner, C. 2010. Ten technology incubators changing the world. *Forbes*, April 16. http://www.forbes.com/2010/04/16/technology-incubators-changing-the-world-entrepreneurs-technology-incubator.html.

Thursby, J. G., and M. C. Thursby. 2003. Enhanced: University licensing and the Bayh-Dole Act. *Science* 301 (5636): 1052.

———. 2005. Faculty patent activity and assignment patterns. Presented at the Roundtable on Engineering Entrepreneurship Research (REER), Georgia Institute of Technology, Atlanta, December.

*Times Higher Education*. 1999. Cambridge nets pounds 100m in research income. Editorial January 4.

———. 2013. *World university rankings*. http://www.thes.co.uk/worldrankings/.

Tolbert, P. S., and L. G. Zucker. 1996. The institutionalization of institutional theory. In *Handbook of organizational studies*, edited by S. R. Clegg, C. Hardy and W. Nord, 175–90. Thousand Oaks, CA: Sage.

Tushman, M. L., and E. Romanelli. 1985. Organizational evolution: A metamorphosis model of convergence and reorientation. *Research in Organizational Behavior* 7: 171–222.

UK Science Enterprise Centers. 2005. *About us*. http://www.enterprise.ac.uk.

University of Cambridge. 2005a. *Cambridge adopts new IPR policy*. http://www.admin.cam.ac.uk/reporter/2005-06/weekly/6039/21.html.

———. 2005b. *The university's mission and core values*. http://www.cam.ac.uk/about-the-university/how-the-university-and-colleges-work/the-universitys-mission-and-core-values.

———. 2013. *Cambridge and facts and figures: February*. Cambridge: University of Cambridge.

US Census. 2008. *State & county quick facts.* http://quickfacts.census.gov/qfd/states/09/0952000.html.

US Patent and Trademark Office. 2003. *US colleges and universities utility patent grants, calendar years 1969–2000.* http://www.uspto.gov/web/offices/ac/ido/oeip/taf/univ/univ_toc.htm.

Van de Ven, A. H. 1986. Central problems in management of innovation. *Management Science* 32 (5): 591–607.

Whelan-Berry, K., J. Gordon, and C. R. Haining. 2003. Strengthening organizational change processes. *Journal of Applied Behavioral Science* 39 (2): 186–207.

Wiewel, W., and D. Perry, eds. 2008. *Global universities and urban development: Case studies and analysis.* Cambridge, MA: M. E. Sharpe.

Yale Office of Public Affairs. 2003. *President's welcome.* http://communications.yale.edu/president/speeches/2003/01/29/universities-and-cities-view-new-haven.

Yale University. 2011–12. *Yale University facts and statistics.* http://www.yale.edu/about/facts.html.

———. 2013. *About: History.* http://www.yale.edu/about/history.html.

Yin, R. K. 2002. *Case study research: Design and methods.* 3rd ed. Vol. 5 of *Applied social research methods.* Thousand Oaks, CA: Sage.

Youtie, J., and P. Shapira. 2008. Building and innovation hub: A case study of the transformation of university roles in regional technological and economic development. *Research Policy* 37 (8): 1188–1204.

Zucker, L. G., M. R. Darby, and J. S. Armstrong. 2002. Commercializing knowledge: University science, knowledge capture, and firm performance in biotechnology. *Management Science* 48 (1): 138–53.

Zucker, L. G., M. R. Darby, and Y. Peng. 1998. *Fundamentals or population dynamics and the geographic distribution of US biotechnology enterprises, 1976–1989.* Cambridge, MA: National Bureau of Economic Research.

Page numbers followed by "f" or "t" indicate material in figures or tables.